ideaFest Journal

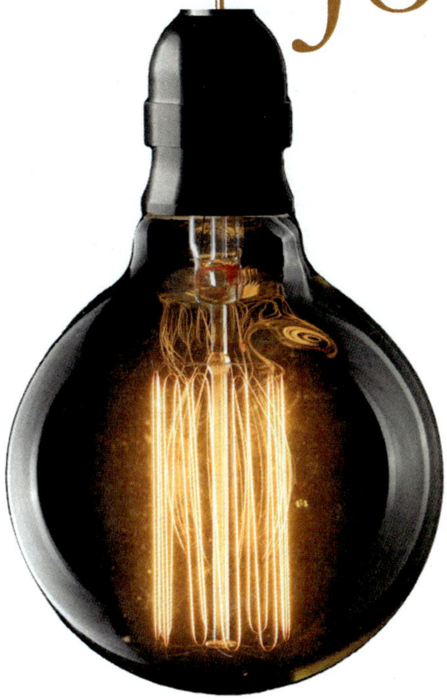

Volume 6
2022

The Interdisciplinary Journal of Creative Works and Research from Cal Poly Humboldt

The Press at Cal Poly Humboldt

Cal Poly Humboldt Library
1 Harpst St.
Arcata, CA 95524-8299

digitalcommons.humboldt.edu

ISBN: 978-1-947112-84-1

ISSN: 2689-1891

Managing Editors: Aaron Laughlin, Kim Sisu
Layout and Design: Kim Sisu, Sarah Godlin

The Press at Cal Poly Humboldt publishes high-quality scholarly, intellectual, and creative works by or in support of our campus community. The press supports the Cal Poly Humboldt mission to improve the human condition and our environment by promoting understanding of social, economic, and environmental issues.

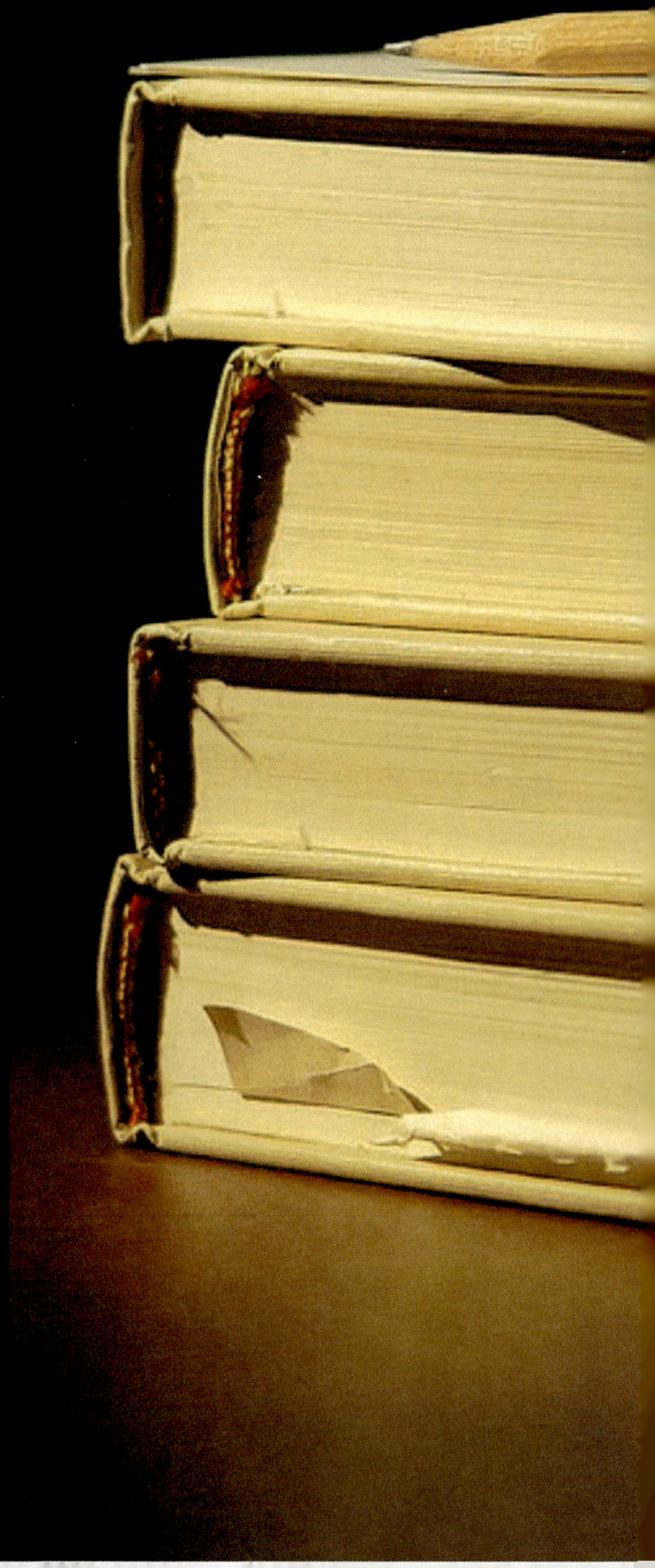

Contents

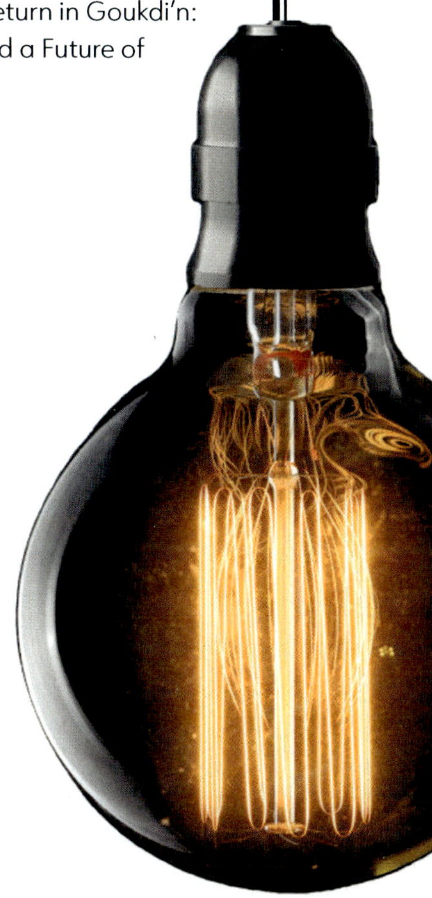

S.O.S.

Sondra P. Schwetman (Cal Poly Humboldt)

Abstract

There are two major bodies of work I generate: one is based on three-dimensional clothing construction, and the other is allegorical figurative work. Both bodies of work display their own poetry. Working with materials such as: Forton MG resin, fibers, bronze, found objects, etc., I feel that possessing knowledge in as many mediums as possible is necessary so that one can achieve a "vision" that is a basis for communication. It is my desire to start a dialogue about women's issues, cultural change, and contemporary miasma. S.O.S. addresses a variety of related issues: bearing witness to our current times, social upheaval, the pandemic, and climate change. Our lives have been changed forever by the last six years as a country and as very divided people. The two figures are in black and white to denote racial and ethnic separations in our culture. They are constructed using body casts in Forton MG: a fiberglass reinforced, non-toxic water-based resin. The figures are also covered in Morse Code, meaning S O S, or ---...--- in opposing colors. They are seated on a conversation chair, an item of furniture made popular in the 19th century. The chair was designed for discreet conversation between two parties, many times for chaste conversation between the amorous. The installation also contains a large number of sewn, silk organza surgical masks beaded with the shape of a medical cross. The masks are laid out in black and white in the pattern of Morse Code. It is my hope to question these changing times, look at them for myself and to highlight, in a visual setting, how our current lives can be discussed through the symbolic.

Keywords: Sculpture, Politics, Social change

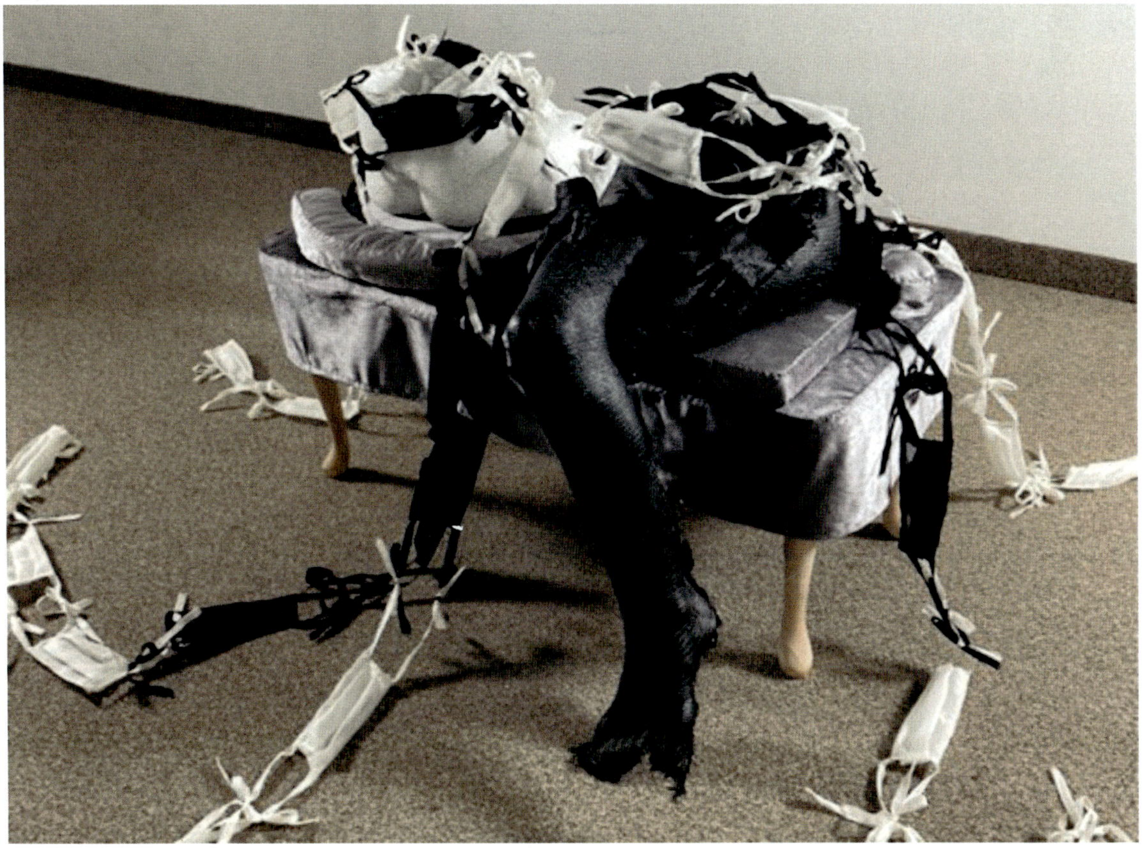

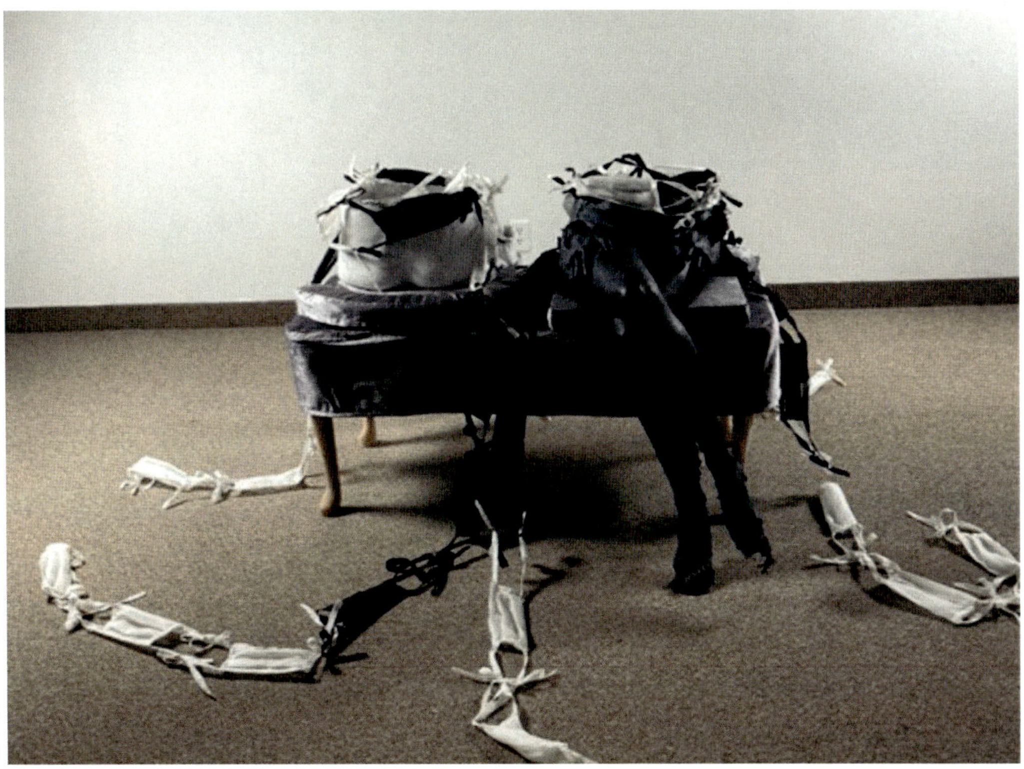

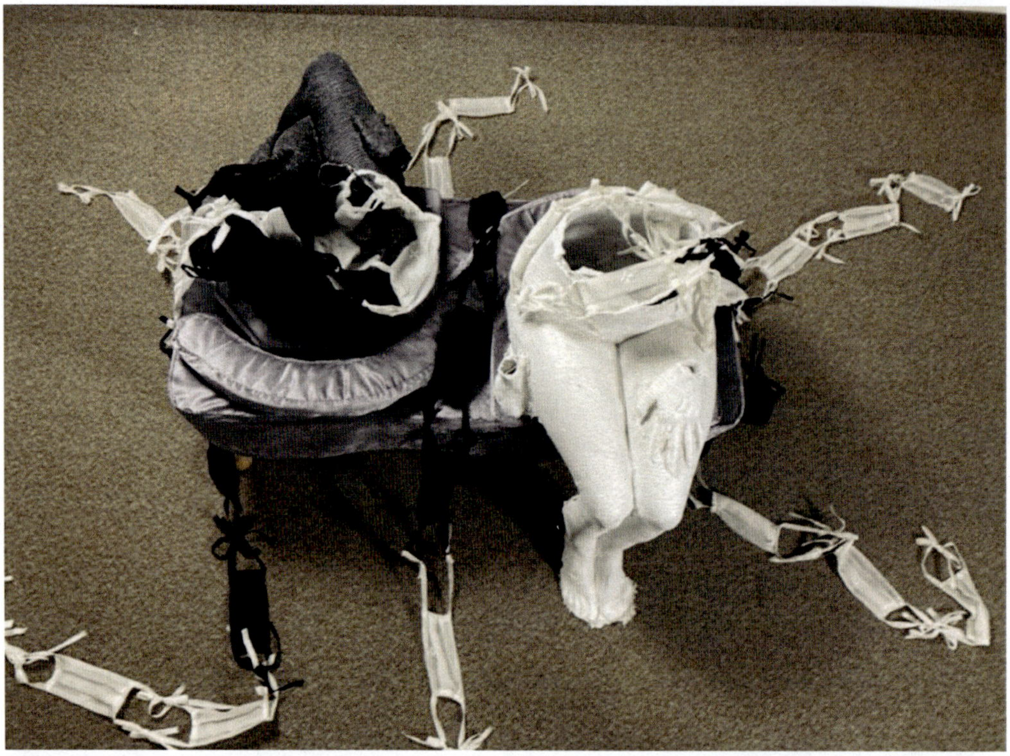

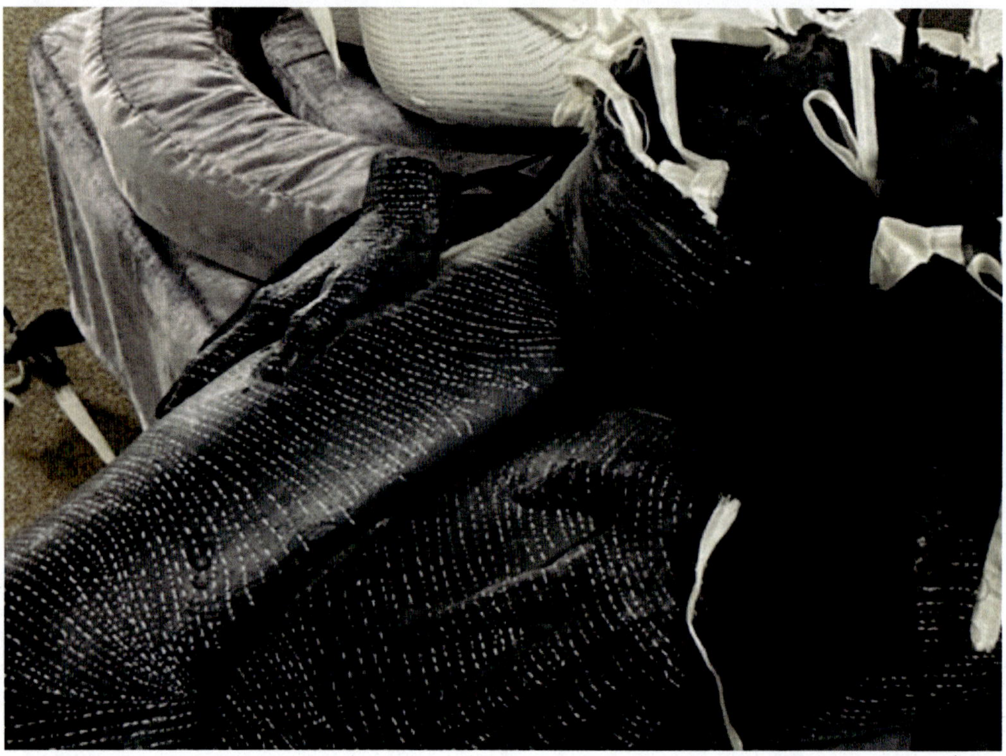

Variation in Coastal Macroinvertebrate Species Diversity on Intertidal Boulders in Trinidad, California

Louis Antonelli (Cal Poly Humboldt), Alexandra Winkler (Cal Poly Humboldt), Theron Taylor (Cal Poly Humboldt), Natalie Greenleaf (Cal Poly Humboldt Alumni)

Keywords: Macroinvertebrate, Diversity, Trinidad, Environment, Intertidal zones, Shannon-Wiener diversity index

Introduction

Coastal environments of Humboldt County are home to a diverse array of marine species. Luffenholtz Beach in Trinidad CA (Figure 1), features rocky intertidal zones and is home to several coastal macroinvertebrate (CM) species (Cimberg, 1975). This region has varying tidal regimes throughout the diverse boulder structures creating unique habitats for aquatic macroinvertebrates. CM species play an important ecological role by cycling nutrients through processing organic matter and making it available for other organisms (Stumpf, et. al, 2009). Boulder habitats on Luffenholtz Beach provide an optimal field of study due to its CM diversity and unique environmental conditions. The focus of this study is on the diversity of CM species on intertidal boulders. We hypothesize that there is a significant difference in the diversity of CM species based on species' habitat elevation on the surface of the intertidal boulders.

Methods

We collected data on October 20th, 2021 at Luffenholtz Beach in Humboldt County, California (41.0401249 N, -124.1200682 W). Tidal charts indicated a high-tide of 6 feet at 11:59 AM and a low-tide of 0.4 feet at 6:09 PM during the time of this study. We took two samples on the eastern side of each of five intertidal boulders (n=10). We counted CM species populations within a one- meter quadrant across two height strata (0-1 meters from base of boulder, and 1-2 meters from the base of boulder). The Shannon-Wiener diversity index (H') was the most appropriate tool to determine the level of diversity per one-meter quadrant measured (Beals, 2000). We then took the averages were then taken at each of the two regions of each boulder. We conducted a two-sample two-tailed t-test in order to determine if there is a significant difference in species diversity between the two regions. Based upon the results of an f-test, the t-test assumed equal variances ($P>0.05$).

Results

We observed eleven total species across all boulders. The resulting data has shown that species in lower regions of the intertidal boulders have a statistically significant higher level of diversity. The p-value from the t-test resulted in a value of 0.041. As shown in Figure 2, the lower region closest to the base of the intertidal boulders (0-1 meters) had on average a 0.452 higher H' diversity index value, than that of the upper region (1-2 meters).

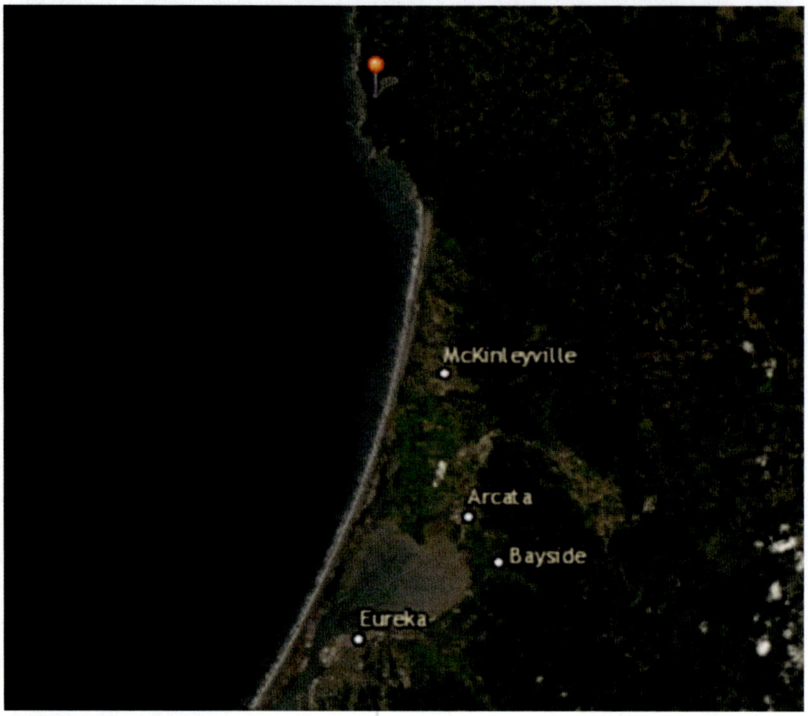

Figure 1. Map of study site in Humboldt County, CA.

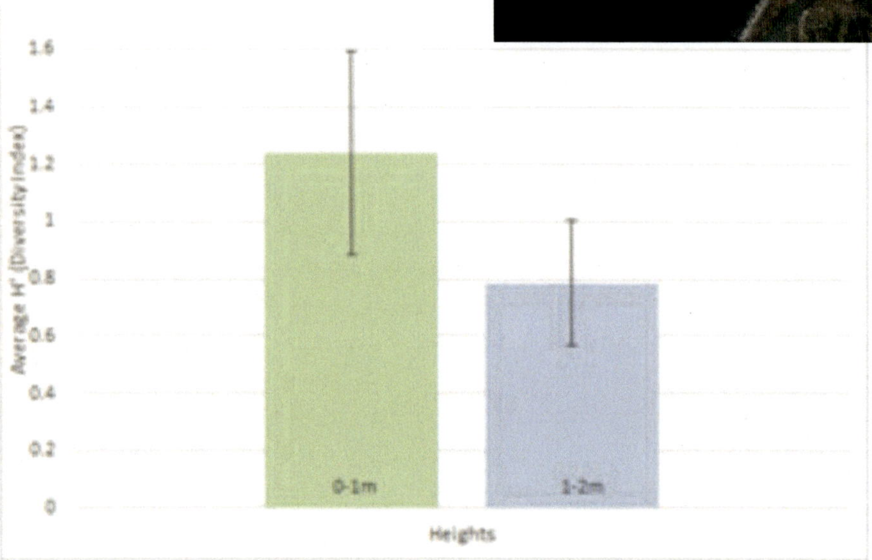

Figure 2. Average Macroinvertebrate Diversity Index (H') at each boulder height (n=10).

Discussion

The results from this study support our alternative hypothesis that there is a significant difference between the diversity of CM species based on their elevation on the intertidal boulders ($P<0.05$). The results of this study may be related to the processes of sessile species adaptation. Sessile CM species can survive on higher elevations of intertidal boulders, while mobile species lack the ability to withstand tidal forces—resulting in higher diversity in lower habitat elevations (Haris, 1990). Sessile species, such as Balanus glandula, use underwater adhesion attaching themselves to desired substrate (Kamino, et al. 2000). Some CM species at higher elevations will seal shut at low tide, reducing exposure to dry conditions and increasing water retention. These processes allow for the scoured substrate to become inhabitable. Variations in tidal range, when sessile species are open and feeding, allow keystone species, such as Pisaster ochraceus, to feed on Balanus glandula in high tide zones; Pisaster ochraceus then return to low tide zones during ebb tide (Lawrence, 2013). The results from this study may be explained by these CM species habitat interactions.

Potential sources of error in this study include a lack of topographic surveys to determine specific rock formations,

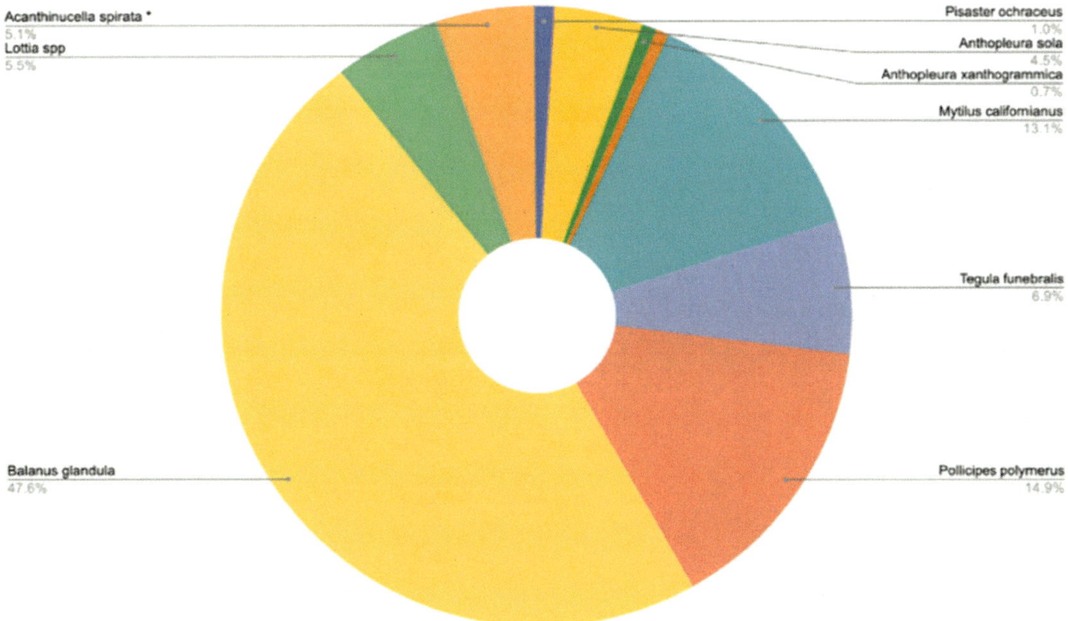

Figure 3. Average CM species population composition across all 5 boulders.

such as differences between boulder fields or benches which promote variation in species diversity along intertidal boulders (Craig, et al. 2017). Rock formation, in connection to elevation, may play a role in the variation of species diversity amongst intertidal boulders (Scrosati, et. al. 2011).

The data from this study indicate that the lower region of the measured intertidal boulder have higher levels of CM species diversity. This provides a better understanding of the habitats for aquatic macroinvertebrates Further research will be needed to better measure and understand CM species diversity in intertidal zones.

References

Beals, M., Gross, L., & Harrell, S. (2000). Diversity indices: Shannon's H and E. The Institute for Environmental Modeling (TIEM), University of Tennessee, USA.

Cimberg, R. (1975, June). Zonation, Species Diversity, and Redevelopment in the Rocky Intertidal Near Trinidad, Northern California. scholarworks.calstate.edu. Retrieved November 4th, 2021, from https://scholarworks.calstate.edu/downloads/vh53x1494?locale=en

Craig, S., Tyburczy, J., Aiello, I., Laucci, R., Kinziger, A., & Raimond, P. (2017, May 31). North Coast Baseline Surveys of Rocky Intertidal Ecosystems. caseagrant.ucsd.edu. Retrieved September 19, 2021, from https://caseagrant.ucsd.edu/sites/default/files/33-Craig- Final.pdf.

Haris, V. (1990). *Sessile animals of the sea shore*. Springer Science & Business Media.

Humboldt County GIS (https://humboldtgov.org/1357/Web-GIS)

Kamino, K., Inoue, K., Maruyama, T., Takamatsu, N., Harayama, S., & Shizuri, Y. (2000, September 1). Barnacle Cement Proteins. *Journal of Biological Chemistry*. Retrieved November 7, 2021, from https://www.jbc.org/article/S0021-9258(19)61519-X/fulltext#:~:text=The%20barnacle%20is%20a%20marine,1.

Lawrence, J. (2013, March). *Starfish: Biology and Ecology of the Asteroidea*. books.google.com. Retrieved December 3, 2021, from https://books.google.com/books?hl=en&lr=&id=WQDvIQ5xR68C&oi=fnd&pg=PA161&dq=Pisaster+ochraceus+tides&ots=EvYeH7s8eP&sig=AzN7IQ9a_uzdRUxRXsswxFPOPi Q#v=onepage&q=Pisaster%20ochraceus%20tides&f=false

Scrosati, R.A., Knox, A.S., Valdivia, N. et al. Species richness and diversity across rocky intertidal elevation gradients in Helgoland: testing predictions from an environmental stress model. Helgol Mar Res 65, 91–102 (2011). https://doi.org/10.1007/s10152-010- 0205-4

Stumpf, S., Valentine-Darby, P., Gwilliam E. (2009). Aquatic Macroinvertebrates - Ecological Role. National Park Service. Retrieved December 1, 2021 from https://www.nps.gov/articles/aquatic-macroinvertebrates-ecological-role.htm

List of Fish Species Present in Galápagos, Ecuador, and California, U.S.A., With Notes on Their Commercial Importance and Conversation Status

Erin J. Hanson (Cal Poly Humboldt), Jose R. Marin Jarrin (Cal Poly Humboldt)

Keywords: California, Commercial fishing, Conservation, Conservation status, Galapagos, Ecuador

Introduction

The purpose of this literature review was to identify fish species that are native to California and the Galápagos Archipelago, and therefore encourage research collaborations between these two important areas.

Methods

The source for the list of Galápagos fish list for this literature review was McCosker and Rosenblatt (2010). Each Galápagos species listed was cross referenced using the Fishbase website (Froese and Pauly 2019) to determine if their native range extended to the California coast. Species that were present in both locations were recorded, as well as their range in California (Southern, Central, and Northern). Existence of fisheries (and type of fishery), IUCN, federal and state protection status was determined from IUCN, US Fish and Wildlife Service and California Department of Fish and Wildlife websites. Regions within California were defined as follows: Southern California, (U.S./Mexico border to Point Conception), Central California (from Point Conception to Pigeon Point), and Northern California, (Pigeon Point to the California/Oregon border).

Results

Of the 550 species of fish found in Galápagos, 100 are also found in California, and 77 of these 100 are the subject of a commercial and/or recreational fishery. These species belong to 52 families, of which the Scombridae (8 species) and Carangidae (7 species) had the most representatives. All 100 species are found in Southern California, but only 58 in Central California and 53 in Northern California. Only one species, Rhincodon typus, is fully protected from all forms of fishing by federal or state action in the U.S. However, according to the IUCN red list of threatened species, two are critically endangered, four are endangered, seven are vulnerable, and four are near-threatened. For the remaining species, 70 are of least concern, eight are data deficient, and five were not included in the IUCN list.

Table 1. List of fish species present in Galápagos and California. If the species is present in Southern, Central or Northern California, existence and type of fishery and IUCN status, and presence or absence of full protection from fishing via the State or Federal Government. Fishery types are described as follows: Small (artisanal fisheries), Large (Commercial), Subsistence, and Recreational. ICUN statuses are listed as follows: CR = Critically Endangered, EN = Endangered, VU = Vulnerable, NT = Near Threatened, LC = Least Concern, DD = Data deficient, NI = Not included in IUCN list.

Common Name	Family	Present up to	Fishery	IUCN
Oceanic whitetip shark	Carcharhinidae	Northern	Large, Recreational	CR
Scalloped hammerhead	Sphyrnidae	Northern	Large, Recreational	CR
Diamond stingray	Dasyatidae	Southern	None	DD
Whiptail stingray	Dasyatidae	Southern	Small	DD
Prickly shark	Echinorhinidae	Northern	Small	DD
Machete	Elopidae	Southern	Small, Recreational	DD
Black marlin	Istiophoridae	Northern	Large, Recreational	DD
Shortbill spearfish	Istiophoridae	Northern	Small, Recreational	DD
Longnose catshark	Pentanchidae	Central	Subsistence	DD
Broomtail grouper	Serranide	Southern	Small, Recreational	DD
Pelagic thresher	Alopiidae	Southern	Recreational	EN
Shortfin mako	Lamnidae	Northern	Small, Recreational	EN
Chilean devil ray	Mobulidae	Northern	Subsistence	EN
Whale shark	Rhincodontidae	Northern	Large, Small, Subsistence	EN
Longsnouted lancetfish	Alepisauridae	Northern	None	LC
Spiny skate	Arhynchobatidae	Northern	None	LC
Finescale triggerfish	Balistidae	Southern	Large	LC
Redtail triggerfish	Balistidae	Southern	Large	LC
Oceanic pufferfish	Balistidae	Southern	Large, Recreational	LC
Californian needlefish	Belonidae	Southern	Small	LC
N/A	Bythitidae	Northern	None	LC
Purple brotula	Bythitidae	Southern	None	LC
Green jack	Carangidae	Central	Large	LC
Amberstripe scad	Carangidae	Northern	Large	LC
Pacific crevalle jack	Carangidae	Central	Large, Recreational	LC
Peruvian moonfish	Carangidae	Southern	Small	LC
Gafftopsail pomapno	Carangidae	Southern	Small, Recreational	LC
Yellowtail amberjack	Carangidae	Northern	Small, Subsistence, Recreational	LC
Scythedmarked butterflyfish	Chaetodontidae	Southern	Small	LC
Milkfish	Chanidae	Southern	Large, Small, Recreational	LC
South American pichard	Clupeidae	Northern	Large	LC
Deepwater conger	Congridae	Northern	None	LC
Common dolphinfish	Coryphaeidae	Northern	Large, Small, Recreational	LC

Common Name	Family	Present up to	Fishery	IUCN
Pompano dolphinfish	Coryphaeidae	Northern	Small, Recreational	LC
Hardtail conger	Crongridae	Southern	None	LC
Pelagic stingray	Dasyatidae	Northern	Small	LC
Slender suckerfish	Echeneidae	Northern	None	LC
Speafish ramora	Echeneidae	Northern	None	LC
Marlin sucker	Echeneidae	Southern	None	LC
Shark sucker	Echeneidae	Southern	None	LC
White suckerfish	Echeneidae	Southern	None	LC
Sharpchin flyingfish	Exocoetidae	Northern	Large	LC
Black wing flyingfish	Exocoetidae	Northern	Large	LC
Spotfin flyingfish	Exocoetidae	Northern	None	LC
Pacific cornetfish	Fistulariidae	Southern	Subsistence	LC
Snake mackerel	Gempylidae	Northern	Small	LC
Black snake mackerel	Gempylidae	Northern	Subsistence	LC
Graceful mojarra	Gerreidae	Southern	Small	LC
Longfin halfbeak	Hemiramphidae	Southern	Large	LC
Pacific silverstripe halffbeak	Hemiramphidae	Southern	None	LC
Indo-Pacific swordfish	Istiophoridae	Northern	Large, Recreational	LC
Yellow snapper	Lutjanidae	Southern	Large, Small	LC
Luvar	Luvaridae	Northern	Small	LC
Abyssal grenadier	Macrouridae	Northern	None	LC
Ocean whitefish	Malacanthidae	Central	Large, Recreational	LC
Slender sunfish	Molidae	Northern	None	LC
Shaprtail mola	Molidae	Northern	Subsistence	LC
Flathead grey mullet	Mugilidae	Northern	Large, Small, Subsistence, Recreational	LC
Mexican goatfish	Mullidae	Southern	Small	LC
N/A	Ogcocephalidae	Northern	None	LC
N/A	Ophidiidae	Northern	None	LC
N/A	Ophidiidae	Northern	None	LC
Longfin bullseye	Priacanthidae	Central	Large	LC
Popeye catalufa	Priacanthidae	Southern	None	LC
Wahoo	Scombridae	Northern	Large, Recreational	LC
Frigate tuna	Scombridae	Northern	Large, Recreational	LC
Skipjack tuna	Scombridae	Northern	Large, Recreational	LC
Black skipjack	Scombridae	Southern	Small, Recreational	LC
Pacific sierra	Scombridae	Southern	Small, Recreational	LC
N/A	Scorpaenidae	Southern	None	LC
N/A	Scorpaenidae	Southern	Subsistence, Recreational	LC
Rainbowbass	Serranide	Southern	None	LC
Threadfin base	Serranide	Southern	None	LC

Common Name	Family	Present up to	Fishery	IUCN
Spotted grouper	Serranide	Southern	Small, Recreational	LC
Star studded grouper	Serranide	Southern	Subsistence	LC
Midwater scorpionfish	Setarchidae	Northern	Small	LC
Pacific porgy	Sparidae	Southern	Large	LC
Bonnethead	Sphyrnidae	Southern	Large, Recreational	LC
Longspined porcupinefish	Tetraodontidae	Southern	Small	LC
Scalloped ribbonfish	Trachiperidae	Northern	None	LC
Lumptail searobin	Triglidae	Southern	Small, Recreational	LC
Smooth stargazer	Uranoscopidae	Southern	Small	LC
Swordfish	Xiphiidae	Northern	Large, Recreational	LC
Paloma pompano	Carangidae	Southern	Small	LC
Ribbon halfbeak	Hemiramphidae	Southern	Small	NI
Striped marlin	Istiophoridae	Northern	Large, Recreational	NI
California grenadier	Macrouridae	Northern	None	NI
Roosterfish	Nematistiidae	Southern	Small, Recreational	NI
Dogface witch eel	Nettastomatidae	Northern	None	NI
Tiger shark	Carcharhinidae	Southern	Large, Recreational	NT
Blue shark	Carcharhinidae	Northern	Large, Small, Recreational	NT
Albacore	Scombridae	Northern	Large, Recreational	NT
Yellowfin tuna	Scombridae	Northern	Large, Recreational	NT
Bigeye thresher	Alopiidae	Northern	Recreational	VU
Silky shark	Carcharhinidae	Northern	Large	VU
Sandbar shark	Carcharhinidae	Northern	Large, Recreational	VU
Oceansunfish	Molidae	Northern	Small	VU
Smalltooth sand tiger	Odontaspididae	Northern	Large	VU
Bigeye tuna	Scombridae	Northern	Large, Recreational	VU
Smooth hammerhead	Sphyrnidae	Northern	Large, Recreational	VU

References

Froese, R., and D. Pauly. 2019. FishBase. World Wide Web electronic publication. Accessed on 7/10/2019

McCosker, J. E., and R. H. Rosenblatt. 2010. The Fishes of the Galápagos Archipelago: An Update. *Proceedings of the California Academy of Sciences 61*(11):167–195.

Websites

Fishbase: www.fishbase.se

IUCN: www.iucnredlist.org

USFW:https://ecos.fws.gov/ecp/listedSpecies/speciesListingsByTaxGroupPage?statusCategory=Listed&groupName=Vertebrate%20Animals&total=400

CDFW: http://nrm.dfg.ca.gov/FileHandler.ashx?DocumentID=109405&inline

Highlighting the Disconnect Between Legislation and Sustainable Cannabis

Johnathon A. Macias

Abstract

Current legislation takes little regard for two major issues challenging cannabis. The carbon footprint and legislation of cannabis are looked at where it is realized that neither of the two issues is sustainable long-term. Solutions that require social responsibility from legislative reform to preserve the culture and industry are delved into.

Keywords: Environmental, Cannabis, Environmental impact of cannabis, Sustainable cultivation, Cannabis policy

Introduction

California classifies cannabis as an agricultural product, yet cannabis still wrestles with "its negative stigma borne out of racial animus" (White, 2012, p.75) that leads to cannabis being abused in terms of accessibility and regulation. Post-prohibition cannabis currently faces two major challenges. The first is misdirected policies surrounding the crop itself which propels the second issue, inefficient production with its capacity for an intensive ecological footprint contributing to the ever-looming threat of climate change. Current legislation holds little regard for either issue leaving farmers and consumers in the dark. Legalization was supposed to benefit the cannabis industry, not shift the burden of knowledge and social responsibility onto the community. With nationwide legalization anticipated in the Marijuana Opportunity Reinvestment & Expungement Act introduced in 2021, it is imperative that policymakers be kept current on the needs of the industry and craft policy to meet the needs of future generations.

Legislation's short-term goals of excessive regulation has pivoted cannabis towards a monopolized market where certain cultivation activities are supported, while other standard practices are delegitimized, (Bodwitch, 2019) deterring some farmers feeling unsupported and reluctant to join the legal market. Cannabis policy should be reformed to assist farmers by streamlining a transparent license process for cannabis with reduced costs of regulation in order to diversify and strengthen the cannabis market.

Environmental Impact

Cannabis operations have the potential for a massive environmental impact that needs to be acknowledged. This environmental damage is exponentially increased in specifically large-scale indoor farms and illicit growers, and if the industry is to thrive into the future it needs to learn to grow in a more environmentally friendly way. The massive amount of energy is attributed to the constant climate control mechanics that indoor growth requires such as HVAC, climate control, and most notably, high-intensity lighting systems.

A large portion of cannabis growing is conducted off-grid; therefore, it can be difficult to precisely measure energy input for indoor cannabis farms. However, a study conducted in 2021 attempted to quantify the greenhouse gas emitted from growing cannabis inside commercial grows, which can vary based on location and growing style but is estimated to range anywhere between "2,283 to 5,184 kg CO2-equivalent per kg of dried flower" (Summers, 2021, p.644). This estimation translates to each pound of cannabis grown in high-intensity conditions emitting at least 1,000lbs of carbon emissions. Furthermore, the 2018 Cannabis Energy Report compiled data from self-reported cultivators and used them as a representative sample to estimate cannabis cultivation energy in the U.S., which was predicted at 1.1 million megawatt-hours (MWh) annually—the equivalent of 92,500 homes (New Frontier). These numbers are problematic because "only about a third of growers identify the environment as a concern" (Bodwitch, 2019, p.181). Without guidance from legislation, the pressure of establishing growing ethics is placed on the cannabis community by holding each other accountable for what is "safe" and what are "harmful" growing practices for the environment.

"Cannabis production is not intrinsically polluting, but engages in inefficient production," (Mills, 2020) and to set a standard for cultivation, the National Cannabis Industry Association put out a highly revered report on the best sustainable management practices for cannabis in 2020 (National Cannabis Industry Association). Understandably, not every cultivator is open to sharing their privatized growing methods, but consumers should be able to ask if these sustainable management practices are incorporated into their farmer's growing process.

Current Legalization Flaws

If well designed, "regulations on agriculture can mitigate impacts on natural landscapes and ecosystems" (Wang, et al., 2017, p.495), but the federal classification of cannabis as a controlled substance comes with federal regulations that catches California and its cannabis farmers in a puzzling interagency battle. Even current research hoping to divulge public health concerns have to dance between red tape just to have the ability to research cannabis.

The current federal framework of policies that monitor the distribution of cannabis disproportionately imposes hardships and inaccessibility onto small-scale farms while simultaneously dismissing the energy intensity of large, indoor, corporate-owned farms. By failing to recognize the true value of excluding small farms from cannabis, poorly crafted policy divides farmers economically by determining their ability to pay the hefty licensing fees that are consistently perceived as "[exclusive] of small growers, undermining economies in rural communities" (Bodwitch, 2019, p.177). An anonymous survey conducted among growers throughout California identified "70% of noncompliance rates being attributed to costs from the multiagency licensing system and 37% struggled with the regulation inconsistencies" (Bowditch, 2019, p.181).

This expensive and complex process is also to manipulation by large cannabis corporation. "Prop 64 was explicitly aimed to build on and support small farms which were supposed to be the backbone of the cannabis cultivation center... but the state undercut small farms by allowing growers to stack permits letting big canna establish larger operations, to get in the game early," (Polson, 2020, p.1) pushing small farms out of the market. Stacked licenses by large operators make up a majority of capacity limits for county ordinances so licenses become even further inaccessible. Ostracizing farmers not only results in a loss in tax contribution but becomes a significant problem when compared to the failure of education about safer and more sustainable growing practices. Policymakers should be more embracing of small farmers because when excluded, farmers have zero incentive to properly dispose of their waste and limited use of testing for pesticides. Over time this waste leaches into the nearest stream resulting in contaminated water that is detrimental to wildlife (Wengert, 2021) and leaves an unnecessary destructive wake in the name of cannabis. Illicit farmers are framed as the perceived villains when the solution of inclusion rests on policy makers hands than their own.

Misdirected Policies

Coined as "the most ubiquitous form of drug taxation" (Valdia, 2015, p.766), the average cost of cannabis from seed sale is widely expensive as legislation misguidedly forces it through several fees before it can begin to be profitable. Cannabis is taxed more heavily when compared to other products. In late 2020 the California legislatures office attempted to compare the tax rates of cannabis to other substances of intoxication and a quarter gram joint was comparable to three shots of liquor (Kerstein 2). A reform to these laws would not only make more rational sense when compared to any other product with a medicinal capacity but would also ease regulations and diversify the cannabis market.

The taxes revenue from cannabis in the state of California as of November 2021 was $3.12 billion (CDTFA, Cannabis Tax Revenues for the Third Quarter of 2021) but these combined fees associated with permits, licensing, and applications end up being too costly for farmers, in fact, "many farms fear that the increased regulatory cost associated with formalization will force them to either shut down or remain on the black market" (Wagner, et al., 2018). Navigating through permits, licensing, and applications, farmers must transform themselves into lawyers before their product can hit the legal market. Some of these regulations such as specific zoning of cannabis plants in relation to other agricultural products are impractical for small-scale operations, making it hard for family farms without financial capital to thrive and risks their profits stop from recycling into the community. Tax revenue is undeniably beneficial for the state but not at the cost of suffocating mom-and-pop farms that are not backed up by a sleuth of investors. Because of issues such as this, the challenge facing cannabis resides in the legislation itself.

Another prime example of a misguided approach that needs to be changed is awarding subsidiaries to large corporate farms for reducing their electricity usage indoors instead of saving recognition for truly remarkable achievements, such as carbon-neutral farms. Encouraging low emission processes is a step in the right direction but can allow for large corporate farms to take advantage of low discounted rate emissions while zero-emissions farms are hardly compensated for their progress towards a sustainable future. If incentives were provided towards achieving zero emissions, farms would strive for a lower emission standard. The proper resources such as financial incentives should be invested in the hands of those reinforcing sustainable thinking, not greenwashing hands looking for cheaper rates.

What Legislation can do

Federal legalization has the capacity to reduce problems associated with the illicit market if crafted more inclusively. Policymakers' rules influence everyday activities, and they need to comprehend that an issue such as a labyrinthine licensing process, prevents farmers from joining the legal market, creates so many more issues than it prevents. Cannabis policy should be reformed in a manner that is inclusive of smaller farmers by offering assistance instead of burdening them with complex licensing processes. Lawmakers should focus on the promotion of information revolving around sustainable cultivation and keeping a more informed consumer base for public health. By promoting the benefits of certain practices such as growing in a greenhouse in comparison to growing with "diesel dope." This would help cannabis deviate from unsustainable practices such as growing with diesel powered generators. Reform would keep community members out of the unpredictable cycle of the black market while reducing the environmental impact of cannabis.

Another way legislation could be more transparent would be more definitive labeling standards. Cannabis products can possibly be interpreted as misleading when policy only requires major cannabinoids such as THC and CBD to be identified on the certificate of analysis. This deceives consumers to think that these cannabinoids are the only significant chemicals in cannabis, but terpenes along with minor cannabinoids also play a major role in determining the effects of cannabis. Many consumers of cannabis looking for a non-synthetic entheogen that eases the hardships of everyday life could be misled by the limited required cannabinoid testing. Susceptible to confusion, consumers can be misled by the analysis certificate and medicinal patients may not get the desired effects sought after. For example, if an extremely high THC product is not performing as advertised, consumers could replace this regulated, safe cannabis with alternative substances such as narcotics, which can be addictive and have devastating side effects instead of the natural relaxation that cannabis brings. In fact, cannabis is so well known as a safe alternative for its pain reliving properties, "promising evidence suggests that cannabis may be a powerful and efficacious tool in the alleviation of this [opioid] crisis" (Wiese and Wilson-Poe 2018). Legislation should put out a public health guide to educate consumers with updated information on the public health status of cannabis instead of leaving a limited surgeon general's warning that most consumers glance over.

California has loosely addressed the environmental concern of cannabis by creating a rule that will go into effect in July 2022, which will require "progress towards compliance with the California Environmental Quality Act before issuing or renewing provisional cultivation licenses." (Department of Cannabis Control). This will help the state bring sustainability into the conversation but still falls short in farmers who lack tangible resources to achieve sustainable gardens.

Sustainable Benefits

Sustainable farmers are environmental stewards that are mindful of energy consumption, conscientious of water use, and make an effort to produce wastes and packaging that can be recycled. A favored method to meet these criteria is regenerative farming in greenhouse operations. Regenerative farming promotes environmental health while greenhouses harness the power of the sun. These sun-grown flowers offer many benefits, but marketing portrays sun grown as second-tier to indoors; therefore, consumers associate the bottom shelf with lower quality. This could not be further from the truth. Sun grown flower has a lower environmental impact and more diverse cannabinoid profiles when compared to indoor.

By relying on the sunlight instead of high-intensity lights, a "shift from indoor grows to outdoor grows could reduce greenhouse gas emissions up to 96%" (Summers, 2021, p.648). In addition to reduced energy consumption, outdoor grows to have a more robust cannabinoid profile than indoors, which merely boast higher THC content. An independent study at a farm in Southern Humboldt compared clones of the same plant grown side-by-side, one indoor and one outdoor (Huckleberry Hill Farms). Their analysis, while not from a peer-reviewed, published study, indicates the possibility that sunlight develops more full-bodied cannabinoid profiles than the clone grown under artificial light. With more cannabinoids to produce a full synergist effect, this study alludes to outdoor cannabis to have more wholesome potential effects than indoor. A drawback to outdoor growing would be susceptible to influences such as male pollen spores, pests, and bad weather. However, these drawbacks can be mitigated through growing in the greenhouse, which can be costly upfront but retains immense long-term value for cultivators. Unfortunately, accessibility to resources leaves greenhouse farming unavailable as an option for many farmers.

Conclusion

Cannabis is not sustainable with its current legislature. Sitting contently with tethering policy directly hurts the workforce behind the industry where smaller farmers are coerced into hard decisions to stay out of debt, such as laying off employees or letting their crops die to prevent debt from taxable harvests. Efficient, sustainable, and healthier alternatives for cannabis could be standardized but legislation seems to remain disconnected from the reality of cannabis and continues failing on integrating small farmers. The regulations that govern cannabis should not limit the industry but should direct it towards a more efficient future. Cannabis has been placed in a negative spotlight for too long by prohibition and with the potential to be cultivated with the healing of the body and soul in mind, the stigmas surrounding cannabis need to be better understood within the context of the wellness it can bring, not baseless claims incited by fear.

Carbon pollution from cannabis is low on the list of contributors to global warming, but in an interconnected world, awareness of our role is significant in preventing a global catastrophe. With multiple perspectives considered, the path towards environmentally friendly cannabis will be complex, but increased relationships and knowledge between policymakers and cultivators will ultimately benefit all of California.

Author Bio

Johnathon Macias has been a professional member of the cannabis community for 6 years. This article is fueled by Johnathon's own experience of a family farm excluded by the licensing process. He works to bring awareness of the environmental impact of cannabis in hopes of bringing sustainability to the conversation. Johnathon believes that Cal Poly Humboldt can play a significant role in the region to build a more environmentally and socially accountable cannabis industry within our state.

References

Bodwitch, H., Carah, J., Daane, K. M., Getz C., Grantham, T. E., Hickey, G. M., & Wilson, H. (2019). *California Agriculture*. 73(3), 177-184. "Growers say cannabis legalization excludes small growers, supports illicit markets, undermines local economies"

California Department of Tax and Fee Administration. "Industry & Tax and Fee Guides." *CA Dept of Tax and Fees*, https://www.cdtfa.ca.gov/industry/.

California Department of Tax and Fee Administration. "California Department of Tax and Fee Administration Reports Cannabis Tax Revenues for the Third Quarter of 2021." *CDTFA Reports Cannabis Tax Revenues for the 3rd Quarter of 2021*, Nov. 2021, https://www.cdtfa.ca.gov/news/21-12.htm#:~:text=Since%20January%202018%2C%20total%20cannabis%20tax%20revenue%20to,Regulate%2C%20and%20Tax%20Adult%20Use%20of%20Marijuana%20Act

Cervantes J., *Marijuana Horticulture: The Indoor/Outdoor Medical Grower's Bible*. Van Patten. 2006

Corva, Dominic, and Joshua S. Meisel. T*he Routledge Handbook of Post-Prohibition Cannabis Research*. Routledge, Taylor & Francis Group, 2022.

"Different Ways to Grow Marijuana." Tips on Growing Marijuana. 2018, https://tipsongrowingmarijuana.com/different-ways-to-grow-marijuana/. Accessed 2021.

Dept. of Cannabis Control. "Requirements for Provisional License 2021'" Department of Cannabis Control, State of California, 2021, https://cannabis.ca.gov/wp-content/uploads/sites/2/2021/10/DCC_Provisional-License-Requirements-1.pdf.

Environmental Sustainability in the Cannabis Industry; Impacts, Best Management Practices, and Policy Considerations. The National Cannabis Industry Association, Oct. 2020, https://thecannabisindustry.org/wp-content/uploads/2020/11/NCIA-Environmental-Policy-BMP-October-17-final.pdf. NCIA-Environmental-Policy-BMP-October-17-final.pdf

Hutmacher, Abby. "Exploring Living Soil in Cannabis Grows." PotGuide, 17 Sept. 2021, https://potguide.com/blog/article/exploring-living-soil-in-cannabis-grows/.

Mills, Evan. "Energy Up in Smoke." Google Sites, Energy Associates, 2020, https://sites.google.com/site/millsenergyassociates/topics/energy-up-in-smoke?authuser=0.

Mills, Evan. "The Carbon Footprint of Indoor Cannabis Production." *Energy Policy*, vol. 46, 2012, pp. 58–67., https://doi.org/10.1016/j.enpol.2012.03.023.

Huckleberry Hill Farms. Photo of Terpene Comparision Bar Chart, analyzed by @vitabudz.nyc, @ridgeline_farms, @huckleberryhillfarms. 4 April 2022. https://www.instagram.com/p/Cb7znEELpaA/

Polson, Michael. "Don't Let Big Agriculture Squeeze out Small Cannabis Farms." Berkeley Blog, UC Berkeley, 2020 https://blogs.berkeley.edu/2020/02/18/dont-let-big-agriculture-squeeze-out-small-cannabis-farms/. Accessed 2022.

Rodd, Scott. "Growers Association Drops Lawsuit Charges against State over Permitting LargeScale Cannabis Cultivators." Bizjournals.com, *Silicon Valley Business Journal*, 2019, https://www.bizjournals.com/sanjose/news/2019/01/22/growers-association-drops-lawsuit-againststate.html#:~:text=The%20Sacramentobased%20California%20Growers%20Association%20had%20filed%20the,that%20legalized%20recreational%20use%20of%20marijuana%20in%20California.

Scale Microgrid Solutions and the Resource Innovation Institute. "The 2018 Cannabis Energy Report." New Frontier Data. October 2018. https://newfrontierdata.com/product/2018-cannabis-energy-report/

Staitman, Gabe. "Growers Perspective on Hypothetical Restricted Outdoor Growing." over-the-phone interview. 14 Mar. 2022.

Summers, H.M., Sproul, E. & Quinn, J.C. The greenhouse gas emissions of indoor cannabis production in the United States. *Nat Sustain 4*, 644–650 (2021).https://doi.org/10.1038/s41893-021-00691-w

Toomey, Diane. "The High Environmental Cost of Illicit Marijuana Cultivation." *Yale E360*, 16 July 2015, https://e360.yale.edu/features/the_high_environmental_cost_of_illicit_marijuana_cultivation.

United States, Congress, Legislative Analyst's Office, and Kerstein. "Comparing Taxes on Cannabis to Taxes on Other Products in California." Legislative Analyst's Office, Dec. 2019. https://lao.ca.gov/Publications/Report/4123. Accessed 2022.

Vaida, Jeremy. 2015. "The altered state of American Drug Taxes." T*he Tax Lawyer 68* (4): 766

Wang, Ian J, et al. "Cannabis, an Emerging Agricultural Crop, Leads to Deforestation and Fragmentation."

Frontiers in Ecology and the Environment, vol. 15, no. 9, 2017, pp. 495–501., https://doi.org/10.1002/fee.1634.

Wagner, Liz. "Small California Pot Farmers Struggle to Survive, Worry That Central Coast Growers Are Using Loophole to Skirt Size Restrictions." NBC Bay Area, 26 June 2018, https://www.nbcbayarea.com/news/local/central-coast-pot-growers-exploiting-loophole-to-skirt-size-restrictions-on-grows/186391/.

Wiese, Beth, and Adrianne R. Wilson-Poe. "Emerging Evidence for Cannabis' Role in Opioid Use Disorder." *Cannabis and Cannabinoid Research*, vol. 3, no. 1, 2018, pp. 179–189., https://doi.org/10.1089/can.2018.0022.

Wengert, Greta M., et al. "Distribution of Trespass Cannabis Cultivation and Its Risk to Sensitive Forest Predators in California and Southern Oregon." *PLOS ONE*, vol. 16, no. 9, 2021, https://doi.org/10.1371/journal.pone.0256273.

White, Kenneth Michael, and Mirya R. Holman. "Marijuana Prohibition in California: Racial Prejudice and Selective-Arrests." *Race, Gender & Class*, vol. 19, no. 3/4, 2012, pp. 75–92. JSTOR, http://www.jstor.org/stable/43497489. Accessed 15 Jan. 2022.

Concentration of Heavy Metals in Three Distinct Algae Families From Humboldt County, California

Kodiak E. Miller (Cal Poly Humboldt, Caleb J. Strait (Cal Poly Humboldt), Jacob I. Begorre (Cal Poly Humboldt), Brittney L. Mitchell (Cal Poly Humboldt), Claire P. Till (Cal Poly Humboldt)

Abstract

Anthropogenic impacts on marine environments can impact metal fluxes and concentrations available to marine species. Monitoring these impacts is necessary to better understand the interactions between the biotic and abiotic components of these ecosystems and mitigate the risk posed by harmful toxins introduced by human activities. Biomoniters, like macroscopic algae, are useful indicators that illuminate the bioaccumulation of toxins commonly introduced from anthropogenic activity. With this in mind, the concentrations of heavy metals zinc (Zn), nickel (Ni), and copper (Cu) were analyzed via the assessment of algae (Representatives from *Ulva*, *Mastocarpus*, *Fucus*) in two sites in Humboldt County: Samoa (urbanized) and Petrolia (rural). Flame atomic absorption spectroscopy (FAAS) was used to quantify the concentration of metals in both algae and sedimental substrate, providing both algal metal content and a biota-sediment accumulation factor (BSAF). It was determined that the order of metal concentration followed Zn > Ni ≥ Cu within algae at both locations for all three algae families. This data is consistent with previous studies of algae species as bioindicators of heavy metal contamination (Kangas et al. 1984).

Keywords: Bioaccumulation, Heavy metals, Biogeochemical cycling, Macroalgae

Introduction

Anthropogenic activities have impacted the biogeochemical flux of elements through ecosystems. Heavy metals are elements of great concern that originate from various activities; from transportation to industrial operations or agriculture, industrialization has intensified the rate and abundance of heavy metal introduction into the environment, which has adversely impacted aquatic ecosystems and organisms therein (Parmar and Thakur, 2013; Sardar et al., 2013). Copper (Cu), nickel (Ni), and zinc (Zn) are present in wastewater effluents that contain leachates from mining, pesticide applications, fertilizer applications, paper production, battery manufacturing, electroplating, and other operations that make their way to aquatic ecosystems (Parmar and Thakur, 2013). Cu has been extensively used in biocidal management as an anti-fouling and anticorrosion agent present in ship hull sheathing (Davy, 1824) but also originates from paper and pulp production, petroleum refining, mining, wood preservation, electroplating, and fiber production (Parmar and Thakur, 2013). Ni has been used to refine precious metals like silver, is common in electroplating techniques, and is popular in battery operations and industry (Parmar and Thakur, 2013). Zn is used for battery operations, mining, galvanization, and smelting and is a common byproduct of industrial paints, polymer stabilizers, fossil fuel combustion, and fertilizer and pesticide applications (Parmar and Thakur, 2013).

Whether present in dilute or concentrated quantities, heavy metals like Cu, Ni, and Zn have mobile, recalcitrant, and persistent qualities that can accumulate within individual organisms and subsequently magnify up trophic levels, ultimately leading to sublethal and lethal concentrations that pose health risks to humans and nonhumans alike (Parmar and Thakur, 2013). Cu, Ni, and Zn are common carcinogens that have been found to cause brain and bone damage in humans, histological and morphological alterations in aquatic organism tissues, and impairment in physiological functions such as growth, development, motility, gas exchange, enzyme activity, behavior, and reproduction (Parmar and Thakur, 2013; Sardar et al., 2013). More specifically, Cu is toxic to organisms above thresholds necessary for its use as a micronutrient (Brand, 1986) and is reported to be highly toxic to kidney, liver, lung, and brain tissues (Parmar and Thakur, 2013). Ni is among the most toxic of metals and has been linked to the onset of asthma and dermatitis as well as lung, nose, and bone cancers (Parmar and Thakur, 2013). Zn exposure has been found to cause neurological stress (Parmar and Thakur, 2013) and impair growth and reproduction (Sardar et al., 2013).

Monitoring the flux and concentration of heavy metals throughout coastal regions is vital for the preservation and development of marine species and ecosystems (Bonanno et al., 2020). However, the flux of metals within a coastal, or more specifically, littoral environment, can be difficult to assess due to the mercurial and wide-ranging contribution of point and nonpoint elemental entrances and exits within the ecosystem.

Algae can be utilized as bioindicators of heavy metals and trace element contamination within a water body at a trophic level (Ho, 1990). A bioindicator is a living organism (algae, animals, trees, microbes, etc.) that allows for assessing the overall quality and health of a specific ecosystem and the species within (Holt et al. 2010). Metal accumulation in algae primarily occurs via adsorption to the cell walls, in which the metals can ionically bind or form complexes with ligands on the cell wall.

As photosynthetic producers, algae inhabit a low trophic level, and elevated heavy metals within them create the risk of biomagnification of toxic metals in other biota that interact both directly and indirectly with algae. Algae are particularly valuable as a bioindicator due to their high absorption rate and high capacity of metals at nearly trace level concentrations (Bonanno, 2020). Furthermore, due to their ability to adapt to diverse environmental conditions, macroalgae can survive and act as an effective bioindicator in both non-polluted and contaminated water environments (Ho, 1990). Clarifying how strongly types of algae select for or against specific elements creates a powerful tool for analyzing biogeographic changes in an environment. An algae's elemental interactions can be determined by measuring metal concentrations using methods of spectroscopy such as flame atomic absorption spectroscopy (FAAS).

In order to study bioaccumulation capacity in algae, a comparison of the algae and its environment is necessary. Metal concentrations within seawater samples tend to be very diffuse and fall below FAAS detection limits, making analysis of marine environments difficult. Additionally, in FAAS of dilute analytes, extremely high sodium chloride concentrations can obscure analyte signals, making detection techniques with greater sensitivity necessary. FAAS is viable when measuring concentrations in sediment due to comparably higher metal concentrations and allows comparison of algae metal concentration with their immediate environment.

The biota-sediment accumulation factor (BSAF) acts as a standardized approach to calculating the bioaccumulation of metals within an aquatic species. BSAF is the ratio of metal concentration within an organism's tissue to the concentration within surrounding sediment. An organism with a high BSAF would be a concentrator, while an organism with BSAF <1 would be a deconcentrator. BSAF is specific to species/analyte relationship and its value and shouldn't change from one location to another.

BSAF provides insight into the abiotic to biotic pathways in an ecosystem. Knowing the primary consumers of high BSAF algae could illuminate species that are particularly vulnerable to contamination, possibly resulting in bioconcentration. The concentration of a measured contaminant can be referenced with the known BSAF of an organism to allow an estimation of that contaminant's concentration within the corresponding species of that region.

For this study, three families of macroalgae were assessed for their suitability as bioindicators of heavy metal. The three genera studied included *Ulva*, green algae from the family Ulvaceae; *Fucus*, brown algae from the family Fucaceae; and *Mastocarpus*, red algae from the family Phyllophoraceae. *Ulva* has been utilized as a bioindicator for heavy metals in literature, in part due to the genus's ecological resilience and speed at which it develops across a wide variety of ecosystems (Gaudry et al., 2007). Metal uptake by Phaeophyta has been characterized in the literature as well (Tropin et al., 1997). The goal of this work was to conduct an initial characterization of algae as metal bioaccumulators across two sites in Humboldt County with distinct and varied anthropogenic stressors.

Materials and Methods

The methods are based on: *Distribution of Elements in Marine Algae of Karachi Coast* (Rizvi and Shameel, 2001) and *A Rapid Wet Digestion Method for Plant Analysis* (Pequereul et al., 2015), as described below.

Sample Collection

The algae samples were collected at two sites in Humboldt County, California, in September 2019. Both sample sites were chosen based on the high availability of algae representing the three macroalgae groups selected for study, as well as the contrasting anthropogenic activities contemporarily and historically affecting the sites. The first site was located at the Samoa Boat Ramp (Latitude: 40.771 N, Longitude: -124.214 W) in the interior of the peninsula of Samoa, California (Google, n.d.), and the second sample site was located at the Devil's Gate Cape (Latitude: 40.408 N, Longitude: -124.391 W) within the Mattole watershed near Petrolia, California (Figure 1) (Google, n.d.). Algae were present in excess at both sites, allowing broad sampling. Both the Samoa and Petrolia sampling sites were sampled at low tides of -0.47 and 0.2 ft, respectively (NOAA 2020).

Algae were selected for their large populations present at sample sites. The benefits of a large population among the sample algae are twofold: a larger, more thoroughly dispersed population leads to a larger possible sample size and a more accurate measurement of the metal concentrations at the sample site.

Algae were collected at low tide above the water line but were within the tidal range. At the Samoa site, algae samples were collected from approximately 05:00-06:00 on September 17, 2020, in high humidity and heavy fog conditions at low tide. Petrolia algae samples were collected from 18:00-20:00 on September 19, 2021, in partially cloudy conditions at low tide. Samples were collected and stored in a freezer until processed within 24 hours of collection.

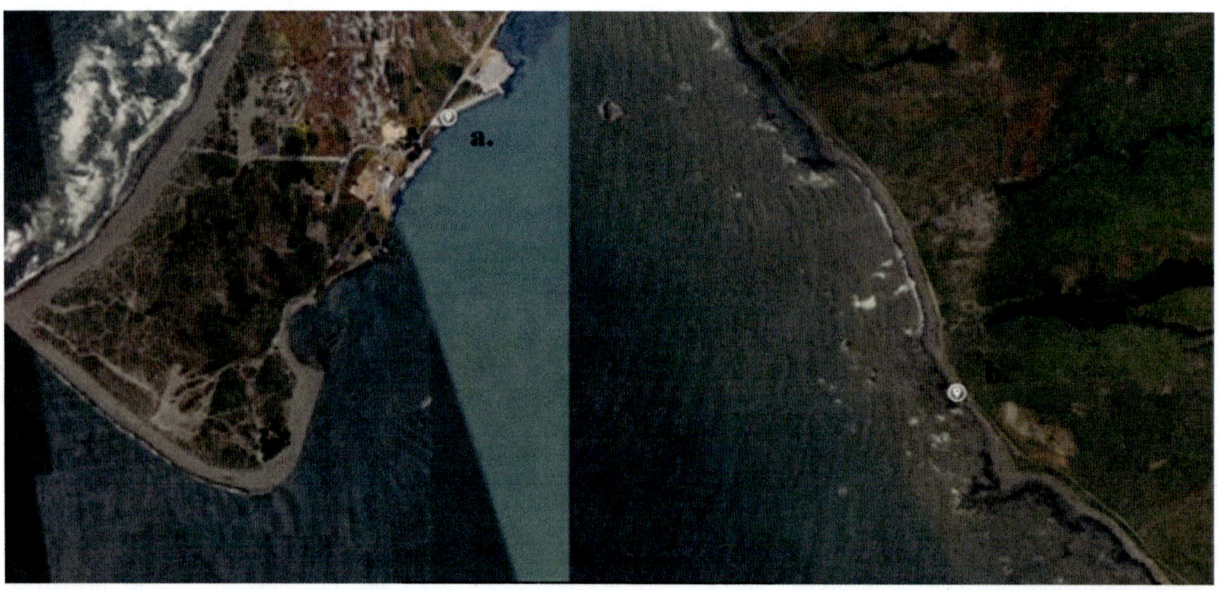

Figure 1. Sample site locations in: a. Samoa, CA and b. Petrolia, CA.

The full thallus (blades, stipe, holdfast) of the algae plant was collected as consistently as possible at both sample sites. Sample individuals were selected semi-randomly on location: efforts were made to spread out the sampling area for a particular algae family to better characterize the overall location. Given the widespread distribution of the algae, samples were selected to maximize the geographic range included in the data set. Algae that appeared unhealthy or colonized with significant epiphytes were omitted from the sampling to avoid metal concentration outliers from atypical algae.

Sample Processing

The algae and sediment samples were placed on ice and cleaned within 6 hours of collection. Prior to cleaning the algae and sediment samples, glassware was treated by acid washing with 10% nitric acid (HNO3). Milli-Q water was used to wash the algal tissue and remove impurities such as sediment and epiphytes. The washed samples were then dried via oven at 65 for approximately 48 hours. (This is a difference from the established methods, which dried the plant matter in the shade.) Any algae samples that were not fully dry after the first 48 hours were placed back into the oven at 65°C to 100ºC until they reached complete desiccation. After drying, algae were ground to a fine powder using a mortar and pestle, which was rinsed with 10% HNO3 and dried with a paper towel between each sample. Samples were then stored in a polyethylene bottle in a cool, dry environment until later analysis.

Algae digestion was based on the method outlined in Jones (1984). All nitric acid and hydrochloric acid solutions were diluted to the proper concentration with Milli-Q water. The dried algae tissue was digested with a dry ashing method at high temperatures in a muffle furnace. Approximately one gram of each algae sample was weighed out within 0.1 mg and added to a porcelain crucible that had been acid washed with 10% nitric acid. The crucibles were placed in the furnace at 500ºC for roughly 6 hours.

After cooling, 10 drops of deionized water and 4 mL of 1:1 diluted nitric acid was added to each crucible containing the ashed algae. The water and nitric acid were evaporated slowly on a hotplate before putting the crucibles back into the muffle furnace for another 2 hours at 500ºC.

Following the previous drying step, 10 mL of 1:1 hydrochloric acid (HCl) was added, and the crucibles were returned to the hotplate to digest for 15 minutes. Following digestion, the remaining solution was vacuum filtered through a Whatman 1 filter paper (11 μm pore size). Once filtered, the solution was transferred into a clean 50 mL volumetric flask, and 5 mL of 1:1 HCl was pipetted into each flask. Each volumetric flask was then diluted to 50 mL with Milli-Q water for later analysis on the flame atomic absorption spectrophotometer.

Standards and Blanks

To prepare the standards, 1000 mg/L stock solutions of each metal were diluted to 50 to 5000 ppb Cu, 75 to 3000 ppb Ni, and 50 to 2000 ppb Zn, each of which was within the linear analytical range for each metal. 0.02 M HCl was added to each standard to keep the metals from adsorbing to the wall of the container. After the standard solutions, additional reagent blanks were created, undergoing the same processing as samples, resulting in solutions containing milliQ, HCl, and any metal ions resulting from lab contamination.

Analysis with Flame Atomic Absorption Spectrometry

Analyst flame atomic absorption spectrometer using an air-acetylene flame and WinLab software. In between each standard or sample, deionized (DI) was aspirated through the FAAS for at least 15 seconds to clear any residual solution or contaminants. Nickel, copper, and zinc absorbances were measured at wavelengths of 232.0, 324.8, and 213.9 nm, respectively.

Data Processing

The Bioconcentration Factor (BCF) compares the concentration of an element in biota to the concentration within the water itself. Metal concentration in plants and algae would be μg/g of moist mass, while metal concentration in water would be μg/mL. Bioconcentration is an indicator of a contaminant entering biota via aqueous routes. Bioconcentration factor was not determined in this study.

$$BCF = \frac{\text{(Metal Concentration (Plant Species)}}{\text{(Metal Concentration (Water)}} \quad (1)$$

The Bioaccumulation Sediment Accumulation Factor (BSAF) is a comparative ratio of the concentration of a contaminant within a biota versus the concentration within the sediment surrounding the biota, with the contaminant measured in μg/g. There are various methods for determining BSAF, which depend on the organism under investigation. Using a method similar to Krivokapic (2021), bioaccumulation factors of the three algae families were calculated as a function the algae concentration divided by the ambient sediment concentration:

$$BSAF = \frac{\text{(Metal Concentration (Plant Species)}}{\text{(Metal Concentration (Sediment)}} \quad (2)$$

(2)

Results and Discussion

The zinc, nickel, and copper concentrations (mg kg-1) within the three collected algae families are shown in Figure 2, along with ambient sediment concentrations. Some of the algae and sediment samples had absorbance values greater

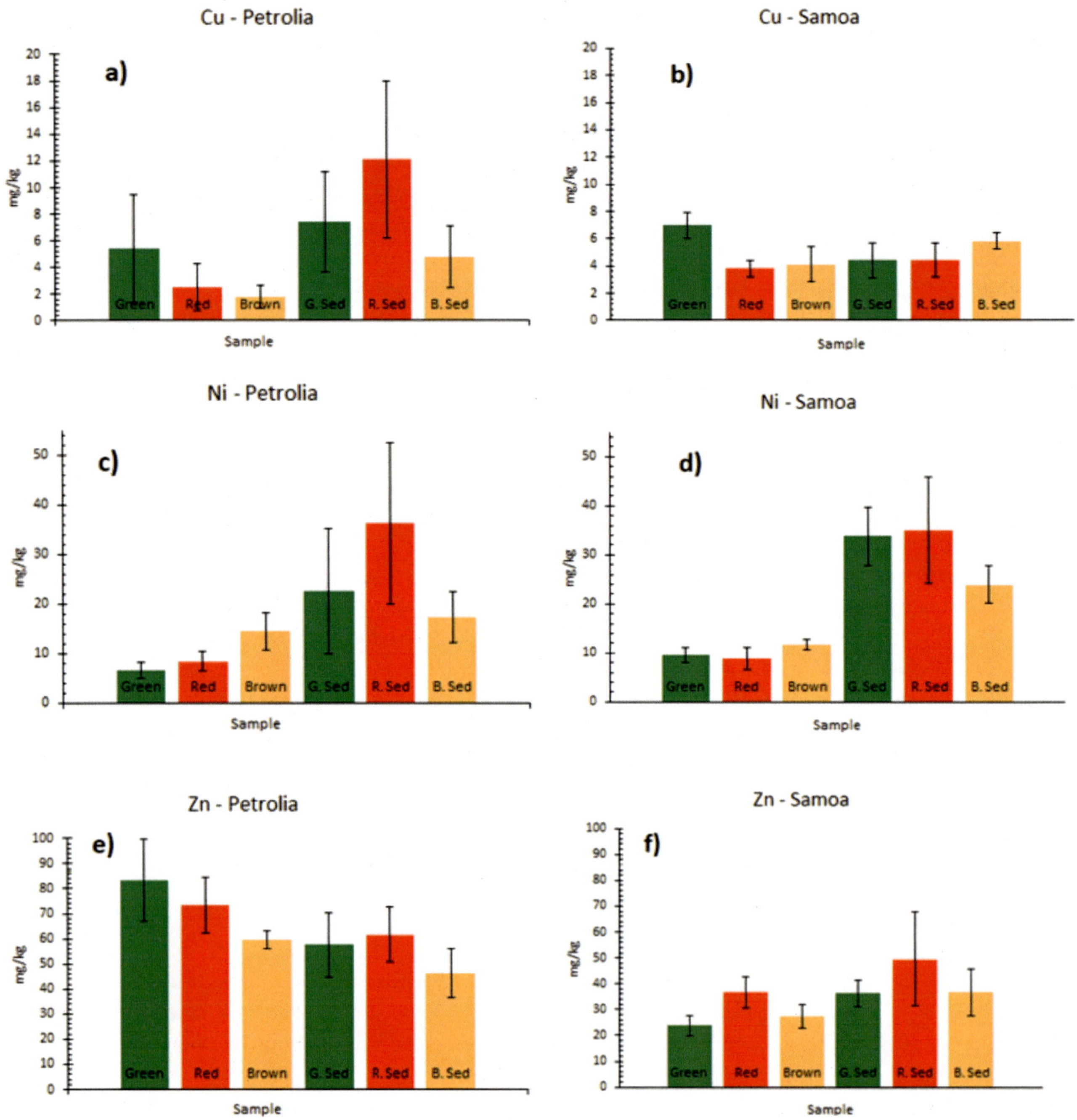

Figure 2. Cu (a, b), Ni (c, d), and Zn (e, f) concentrations in algae and sediment samples in Samoa (left panels) and Petrolia (right panels).

than the limit of linearity or outside the range of the standards; therefore, those concentrations will be underestimated in the final data.

Comparing Metals

The overall trend for the metal concentrations across all three algae families followed the order Zn > Ni ≥ Cu (Figure 2), which is the same trend found for *Ulva* in the Bonnanno et al. (2020) paper. According to Romera et al. (2007), the affinity of the metals to algae biomass is ordered Cu > Ni > Zn (average of six algae species), indicating that copper is the most likely to adsorb to the surface of the algae. Furthermore, the maximum sorption of these three metals to biomass is ordered as Cu > Ni > Zn. Based on these absorption constants, it would be expected that copper has the greatest concentration within the algae families. However, this trend was not observed (Figure 2). It could be that Cu concentrations were relatively low in both sampling locations. This may suggest that anthropogenic sources of Cu are not significantly influencing the concentrations in Samoa and Petrolia, or there are system interactions that impact Cu uptake in algae. Alternatively, there could be nonpoint sources of Zn in both Samoa and Petrolia leading to relatively high Zn concentrations in those locations, which would explain the relatively high Zn concentrations in the algae. Assessment of Humboldt watersheds cites fertilizer and pesticide runoff as concerns (North Coast Regional Water Quality Control Board, 2005), both of which are plausible sources of Zn (Parmar and Thakur, 2013).

All three metals analyzed are utilized as micronutrients and are naturally present biotically and abiotically within the sampled environments. Environmental chemical processes at play are broad and difficult to isolate from one another and can obfuscate anthropogenic impacts on metal concentration within these systems; however, it can be broadly stated that sources that deliver metal ions into the observed system result in increased concentrations.

Comparing Algae Families

For the Petrolia samples, Cu and Zn concentrations follow the same pattern across algae: the highest metal concentrations were found in the green algae, followed by the red algae, and the lowest concentrations in the brown algae (Figure 2). However, nickel followed the opposite trend, with the highest concentrations in the brown algae, followed by the red algae, and then the green algae. This could indicate that the brown algae have a comparatively high nickel requirement than the green and red algae and are preferentially taking it up from the abiotic environment to a higher degree.

Comparing Sediment and Algae Concentrations

The biota-sediment accumulation factor (BSAF) was analyzed between algae and their sediments, and the calculated BSAF values (Equation 2) were averaged between sampling sites (Figure 3).

With regards to nickel, all three families took up less nickel than in their environment, with the brown algae taking

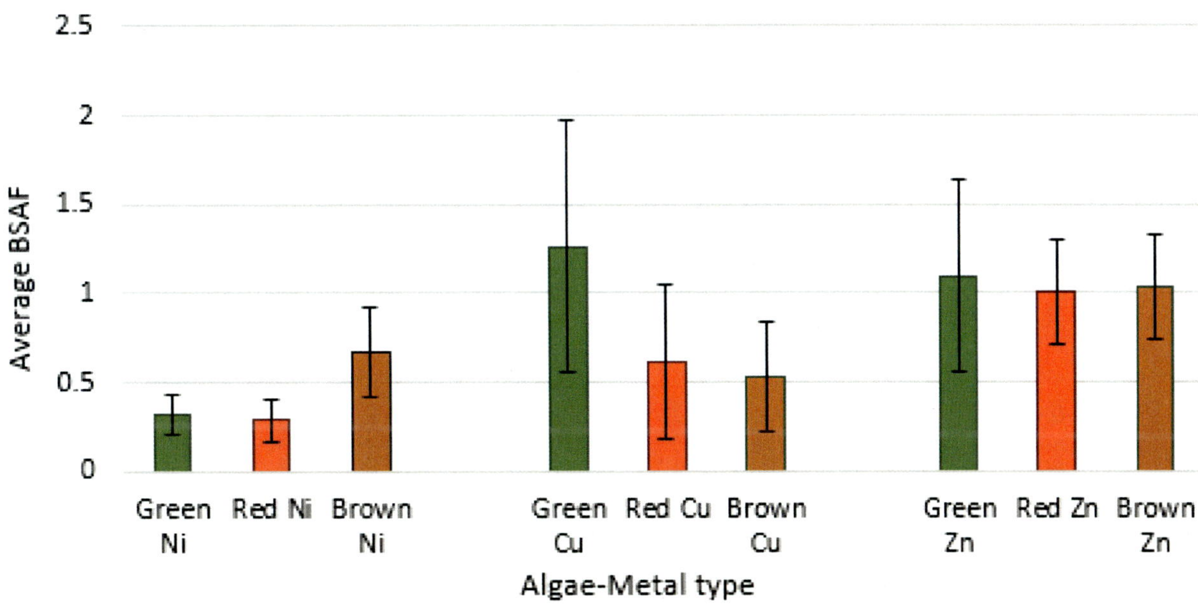

Figure 3. Calculated Petrolia & Samoa Average BSAF values (Zn, Cu, Ni) across algae.

up slightly more than green and red. Green algae acted as a concentrator of copper, while red and brown had a lower copper concentration compared to sediment background. Site-specific BSAF values shown below in figure 4 show a significant disparity in algae BSAF depending on the site. The difference in site BSAF values suggests either environmental factors that influence how a species selects for or against specific nutrients or an issue within the method disproportionately affecting one site over the other, as discussed below.

All three families in Petrolia showed a BSAF of 1 or greater for Zn, but all three families showed less than 1 in Samoa. At the Samoa sampling site, zinc concentrations were significantly lower in the algae samples when compared to Petrolia, leading to a greater BSAF value. Across both sample sites, nickel concentrations within sediment were noticeably greater than the concentrations in the algae samples (Figure 2), leading to relatively low BSAF values (Figure 3). This could be due to the algae having a lower biological need for nickel than the other metals and therefore taking it up less actively. Although published work for metal requirements in marine macroalgae is limited, Twining and Banes (2013) found that most marine phytoplankton required less nickel than zinc, supporting this hypothesis. However, most phytoplankton species required even less copper than nickel.

Comparing Sites

The Samoa and Petrolia sampling sites were chosen as a comparative pair due primarily to their respective urban and rural surroundings. The Samoa sampling site is located within the Humboldt Bay watershed, where approximately 65,000 (59%) of the county population resides (Humboldt County Department of Health & Human Services, 2018). From this disproportionate population density, we expect a greater anthropogenic effect on the water ion composition within the sample site. Land use outside of the urban centers present includes timber production as well as agricultural grazing and dairies. The bay itself is a commercial, deep-water port and a site of significant commercial shellfish production. Due to commercial bay use and urban factors such as US Route 101 that bisects the watershed, greater metal contamination risk is expected. Reported water quality issues that would contribute to metal contamination include sedimentation from tributaries, automotive use, and shipping (North Coast Regional Water Quality Control Board, 2005). The Petrolia sample site located further south is contained within the Mattole watershed. The population within this watershed is much more diffuse, totaling approximately 2,000 people. Aside from 20th century timber harvest creating sedimentation issues via erosion, the Mattole watershed experiences comparatively less commercial and urban anthropogenic elemental agitation. There are no major roadways present within the watershed, and the sample site was exposed to the ocean, suggesting a more abstracted exposure to commercial shipping compared to the Samoa sample site within the bay (North Coast Regional Water Quality Control Board, 2005). The contrasts between sample sites suggest that the Samoa site would have greater anthropogenic disturbance contributing to more significant metal pollution; however, the data did not depict this.

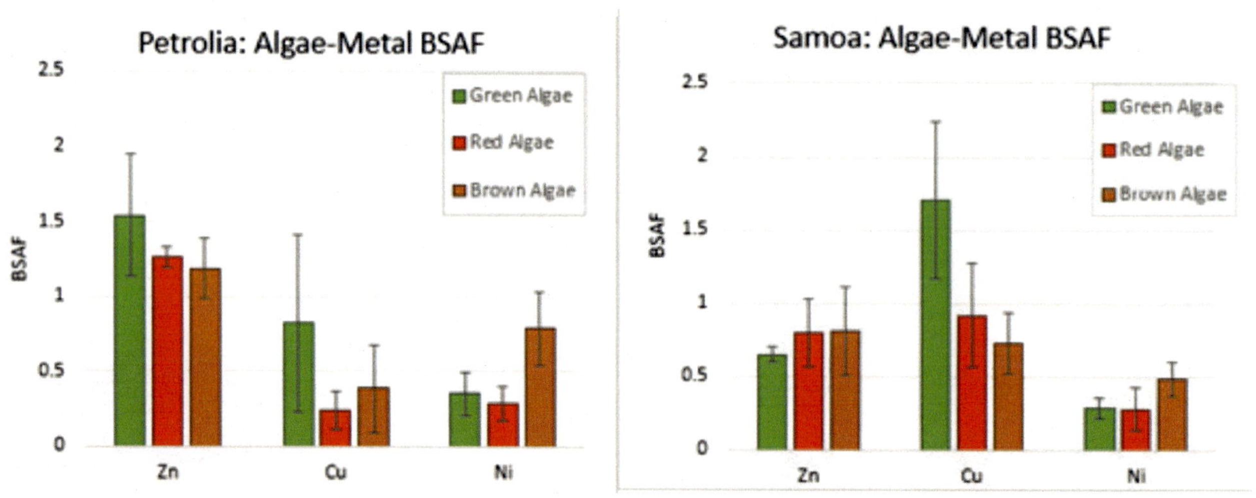

Figure 4. Calculated BSAF values (Zn, Cu, Ni) of the three algae families in Petrolia (left) and Samoa (right).

Comparing biotic and abiotic concentrations of the elements of interest regarding bioaccumulation yields several observable patterns. The sediment samples from Samoa were measured with lower Cu and Zn concentrations than Petrolia. All algae collected at the Samoa and Petrolia sites exhibited lower Ni concentrations than their surrounding sediment samples. Due to the algae exhibiting lower levels of Ni than ambient sediment, it is assumed that these algae do not concentrate Ni. Furthermore, this observation is indicative that sources of Ni are naturally occurring and that the algae is not contaminated with this metal.

Of the three metals measured, Cu had the lowest concentration in both algae and sediment samples at both sites.

Sources of Error

Samples that required additional drying in the initial oven treatment were likely subject to increased decomposition in the period between washing and drying of the samples. The desiccation time had the most significant disparity between families regardless of sample sites, with samples from *Fucus* generally requiring additional drying time compared to both *Ulva* and *Mastocarpus*.

At both sample sites, the full thallus (blades, stipe, holdfast) of the algae plant was collected as consistently as possible. However, the holdfast on several samples was not collected due to its strong retention to the sediment. Given that no particular family of algae was especially prone to this error, the impact on collected data is presumed to be marginal.

During the initial nitric acid evaporation step for Samoa samples, several sample crucibles experienced a loss of sample due to bumping from rapid heating. The effects this had on analyte concentration results cannot be reliably predicted. It was decided that while this is a source of error, it is unlikely that the observed metal uptake results were skewed enough to render them invalid for the purposes of this study.

Significant differences in sediment at both sample sites may have also impacted the mobilization of metals within the samples. Qualitatively, the substrate gathered at the Samoa site could be classified as sand, whereas the grain size for several samples at the Petrolia site was much larger and more pebble-like in quality. Given the discrepancy in surface area between sites, there was concern that the acid treatment would lead to a greater concentration of metals within the same mass of Samoa sediments. A rough assay can be conducted by averaging the metal concentration in Petrolia sediment samples that were noted as being larger in grain size compared to those with smaller-grain substrate. As visualized in figure 5, the results of this averaging show a greater average concentration in sediment samples with smaller grain sizes for all three metal ions analyzed.

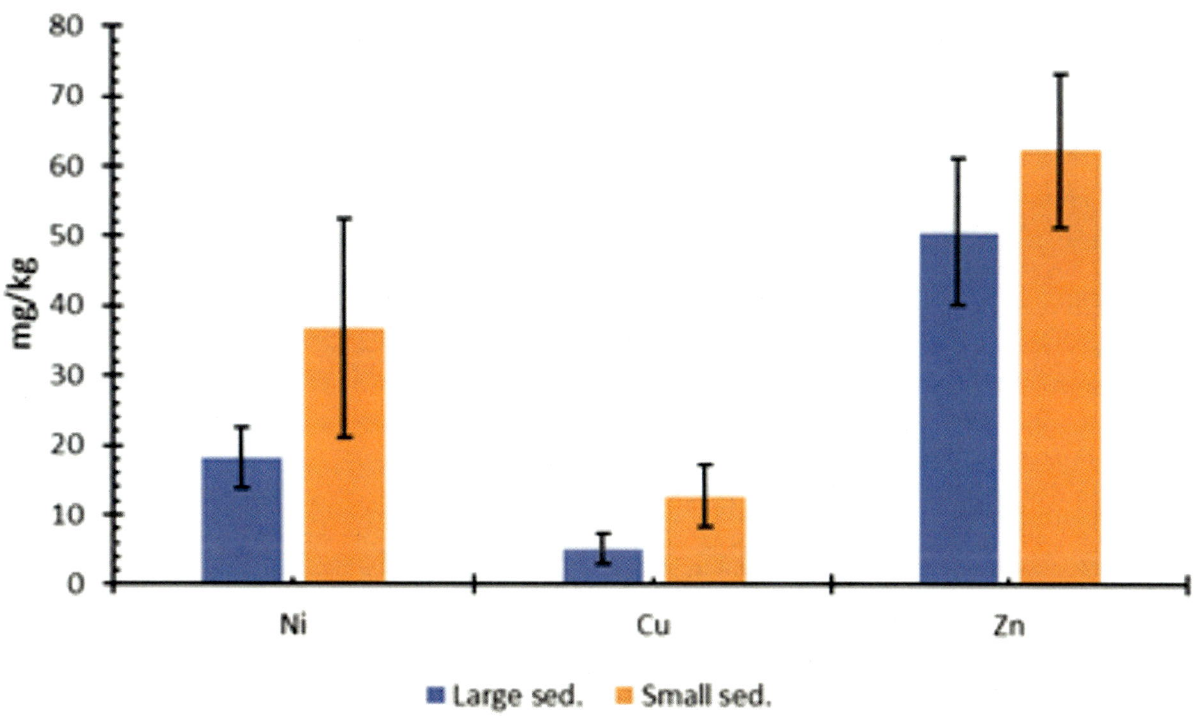

Figure 5. Metal concentration comparison between Petrolia sediment samples marked as being larger in size.

The effect on grain size in the measurement of sediment metal concentration has possible impacts on the resulting BSAF calculation. If larger grain size leads to a reduced quantity of metal during processing, the resulting BSAF of the algae at Petrolia would appear larger as a result of lower apparent metal concentration. Ni and Zn BSAF values are greater at the Petrolia site compared to Samoa; however, all algae had a lower Cu BSAF at Petrolia. If the larger sediment size at Petrolia artificially increased BSAF for all metals, then our conclusions based on the BSAF factor would need to be adjusted: there may be more pressure to select against copper at Petrolia or more pressure to select for copper at Samoa.

Conclusion

Anthropogenic activity has undoubtedly disrupted the concentration and flux of metals and other elements within environmental systems, and determination of species capable of acting as biomonitors is the first step to elucidating these changes. In this exploratory study, three algae genera (*Ulva, Mastocarpus, Fucus*) were compared for their suitability as bioindicators with regards to Cu, Ni, and Zn at urban and rural sites in Humboldt County using FAAS. Results suggest that all algae deconcentrate Ni, *Ulva* concentrates Cu, and all algae neither concentrate nor deconcentrate Zn.

The results of this study do suggest that certain algal families have predilections towards certain metals, but there is significant variability in these results, and more research is needed to have a conclusive understanding of these trends. Furthermore, this study was limited to two locations sampled once each, and a broader dataset would allow broader interpretations and conclusions. We propose the following future investigations to fill the gaps. Cross substrate analysis of water in addition to sediment metal concentrations would clarify how factors such as pH might affect measured concentrations in sediment as well as algae. Future work involving a comparison of sites would benefit from including a broader range of sites to reduce the likelihood of site to site processing differences (e.g., sediment surface area) disproportionately affecting the resulting BSAF of algae analyzed. Future studies could also assess whether there are metal concentration thresholds that a species will accept or reject from surrounding sediment and whether the BSAF of a given algae is static or dependent on its surroundings or its life stage.

References

Davy, H. (1824). XII. Additional experiments and observations on the application of electrical combinations to the preservation of the copper sheathing of ships, and to other purposes. *Philosophical Transactions of the Royal Society of London*, 114, 242–246. https://doi.org/10.1098/rstl.1824.0015

Gaudry, A., Zeroual, S., Gaie-Levrel, F., Moskura, M., Boujrhal, F.-Z., El Moursli, R.C., Guessous, A., Mouradi, A., Givernaud, T., Delmas, R., 2007. Heavy Metals Pollution of the Atlantic Marine Environment by the Moroccan Phosphate Industry, as Observed through their Bioaccumulation in Ulva Lactuca. *Water, Air, & Soil Pollution 178*, 267–285.. doi:10.1007/s11270-006-9196-9

Ho, Y.B., 1990. Ulva lactuca as bioindicator of metal contamination in intertidal waters in Hong Kong. *Hydrobiologia 203*, 73–81.. doi:10.1007/bf00005615

Holt, E. A. & Miller, S. W. (2010) Bioindicators: Using Organisms to Measure Environmental Impacts. *Nature Education Knowledge 3*(10):8

Humboldt County Department of Health & Human Services, Public Health. (2018). 2018 Humboldt County-Community Health Assessment. https://humboldtgov.org/DocumentCenter/View/71701/2018-Community-Health-Assessment-PDF

Jones, J. B. Jr. (1984) Plants. In *Official Methods of Analysis of the Association of Official Analytical Chemists*. Ed. SWilliams. pp 38-64. Association of Official Analytical Chemists, Arlington, Virginia 22209, USA.

Kangas, P., & Autio, H. (1984). Macroalgae as indicators of heavy metal pollution. P*ublications of the Water Research Institute*, National Board of Waters, Finland, 68, 183-189.

Krivokapic, M. (2021). Study on the Evaluation of (Heavy) Metals in Water and Sediment of Skadar Lake (Montenegro), with BCF Assessment and Translocation Ability (TA) by Trapa natans and a Review of SDGs. *Water 2021*, 13(6), 876; https://doi.org/10.3390/w13060876

Kochoni, E., & Fortin, C. (2019). Iron Modulation of Copper Uptake and Toxicity in a Green Alga (Chlamydomonas reinhardtii). *Environmental Science & Technology*, 53(11), 6539–6545. https://doi.org/10.1021/acs.est.9b01369

Klimmek, S., Stan, H.-J., Wilke, A., Bunke, G., & Buchholz, R. (2001). Comparative Analysis of the Biosorption of Cadmium, Lead, Nickel, and Zinc by Algae. *Environmental Science & Technology*, 35(21), 4283–4288. https://doi.org/10.1021/es010063x

National Oceanic and Atmospheric Association. (2020). Tide Predictions at 9418817, Samoa, Humboldt Bay CA. https://tidesandcurrents.noaa.gov/noaatidepredictions.html?id=9418817&units=standard&bdate=20200901&edate=20200930&timezone=LST/LDT&clock=12hour&datum=MLLW&interval=hilo&action=monthlychart

North Coast Regional Water Quality Control Board. (2005, February). Watershed Planning Chapter. https://www.waterboards.ca.gov/northcoast/water_issues/programs/wpc/18humboldtsec2.pdf

North Coast Regional Water Quality Control Board. (2005, February). Watershed Planning Chapter. https://www.waterboards.ca.gov/northcoast/water_issues/programs/wpc/09mattolesec2.pdf

Parmar, M., & Thakur, L. S. (2013). Heavy metal Cu, Ni and Zn: toxicity, health hazards and their removal techniques by low cost adsorbents: a short overview. *International Journal of plant, animal and environmental sciences*, 3(3), 143-157.

Romera, E.; González, F.; Ballester, A.; Blázquez, M. L.; Muñoz, J. A. Comparative Study of Biosorption of Heavy Metals Using Different Types of Algae. *Bioresource Technology 2007*, 98 (17), 3344–3353. https://doi.org/10.1016/j.biortech.2006.09.026.

Sardar, K., Ali, S., Hameed, S., Afzal, S., Fatima, S., Shakoor, M. B., ... & Tauqeer, H. M. (2013). Heavy metals contamination and what are the impacts on living organisms. *Greener Journal of Environmental management and public safety*, 2(4), 172-179.

Tropin, I. V.; Konstantine S. Bourdine. Periodic Changes in Metal Content in Fucus Distichus Thalli under Polar Day Conditions. *Journal of Applied Phycology 1997*, 9, 269–276.

Supplementary Information

Table 1. Samoa samples

	Ni Avg. (mg/kg)	Ni Stdev	Cu Avg. (mg/kg)	Cu Stdev	Zn Avg. (mg/kg)	Zn Stdev
Green	9.57	1.53	6.99	0.95	23.76	3.86
Red	8.86	2.21	3.81	0.58	36.67	6.14
Brown	11.70	1.04	4.13	1.26	27.38	4.65
G. Sed.	33.86	5.95	4.41	1.29	36.24	5.28
R. Sed.	35.10	10.76	4.43	1.23	49.62	18.00
B. Sed.	24.00	3.73	5.85	0.63	36.67	9.24

Table 2. Petrolia sample data

	Ni Avg. (mg/kg)	Ni Stdev	Cu Avg. (mg/kg)	Cu Stdev	Zn Avg. (mg/kg)	Zn Stdev
Green	6.60	1.62	5.44	4.03	83.38	16.42
Red	8.42	2.03	2.53	1.75	73.37	11.24
Brown	14.48	3.78	1.79	0.88	59.62	3.47
G. Sed.	22.53	12.67	7.41	3.76	57.50	12.95
R. Sed.	36.25	16.28	12.08	5.89	61.65	10.87
B. Sed.	17.33	5.02	4.79	2.30	46.47	9.70

Table 3. Petrolia BSAF averages

	Zn	Cu	Ni
Green	1.536	0.820	0.352
Brown	1.184	0.386	0.787
Red	1.261	0.243	0.287

Table 4. Samoa BSAF averages

	Zn	Cu	Ni
Green	0.652	1.701	0.290
Brown	0.809	0.731	0.486
Red	0.798	0.918	0.280

Table 5. Large sediment comparison between Petrolia samples

	Ni	Stdev Ni	Cu	Stdev Cu	Zn	Stdev Zn
Large sed.	18.16	4.31	5.15	2.07	50.69	10.52
Small sed.	36.70	15.71	12.72	4.51	62.30	10.98

Table 6. Algae

	Ni	Cu	Zn
Green	0.320655 ± 0.110562	1.260489 ± 0.708	1.094144 ± 0.53787
Red	0.283493 ± 0.283	0.61117 ± 0.439	1.008847 ± 0.29521
Brown	0.666725 ± 0.249	0.524148 ± 0.306969	1.033881 ± 0.300

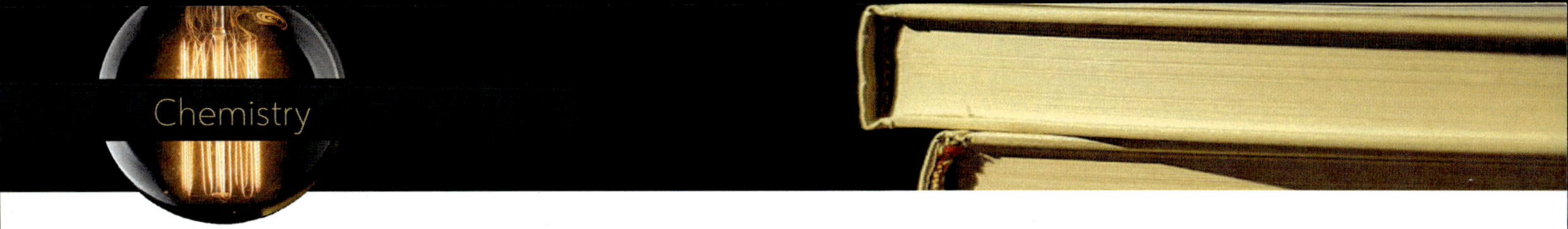

Design of Possible Organic Photovoltaic Compounds and Their Initial Computational Assessment

Albert Ochoa Castillo (Cal Poly Humboldt), Joshua Smith (Cal Poly Humboldt)

Abstract

The excessive use of fossil fuels has surged the need for alternative energy sources, such as solar energy. Organic photovoltaic (OPV) cells can potentially be an alternative to silicon solar cells. There is a large interest in OPV cells because they are strong absorbers and come at a low cost of synthesis. In this study, possible OPV compounds were designed, and theoretical computations were done to assess their efficiency. Density Functional Theory was used to calculate the HOMO-LUMO gap of the compounds designed. Semiconductors such as naphthalene, 1,1'-biphenyl, and α-septithiophene were used as the backbone with electron-withdrawing groups (EWG) attached. Seven of the 26 compounds demonstrated adequate results and four of them will be further pursued to synthesize. Lastly, results suggest there is most likely a correlation between the amount of EWGs attached to the semiconductor and its calculated efficiency.

Keywords: Organic photovoltaic compounds, Thiadiazole, HOMO-LUMO gap

Introduction

In an economy where the primary source of energy comes from fossil fuels, there is an urgent need for alternative energy sources. Solar energy has become increasingly popular due to its feasible transformation into electricity through photovoltaic cells.[1,2] A typical photovoltaic cell consists of two silicon semiconductor layers where one layer contains positively charged (p-type) holes and the other contains negatively charged (n-type) holes. When the sun hits the cell, photons push electrons out of their place and create more holes in the n-type layer. In the p-n junction—where the two layers touch—an electric field stops electrons and holes from moving between layers. However, the layers are connected in a circuit, thus the electric field pushes electrons through the circuit, and this creates a current.[1] For example, silicon-based photovoltaic cells are widely used because of their high efficiency and their low cost. Although silicon-based photovoltaic compounds have a favorable band gap of about 1.1 eV, they do not absorb visible light well due to the indirect nature of the gap.[1] Having a band gap of 1.1 eV would mean silicon cells can theoretically absorb 77% of incoming solar photons.[3] However, silicon cells are often difficult to recycle after they are retired and leave a large carbon footprint after production.[4] This is why there is a large interest in organic photovoltaic (OPV) compounds, as OPV cells are strong absorbers and come at a low cost of production (synthesizing).[3] Because OPV devices use a much smaller quantity of material, they should also have a smaller carbon footprint than traditional solar cells.

An OPV bilayer device is composed of an OPV compound (the active layer) that is sandwiched between a conductive electrode and a transparent conductive substrate.[3] Everything is then covered with glass or plastic. Although the architecture of an OPV device is critical, there is a focus on the active layer because it is the main OPV compound that absorbs sunlight.

In general, OPV compounds contain a conjugated polymer as the backbone.[3] Electron-withdrawing groups (EWG) and electron-donating groups (EDG) are then attached to the polymer to increase its efficiency. Some of the most efficient semiconductors used in OPV compounds include pentacene, perylene, buckminsterfullerene, sexithiophene, poly(triarylamine), phthalocyanine, poly(3-hexylthiophene), and poly(3-vinylthiophene).[3] Productive OPV compounds are mostly derived from the semiconductors listed above. It is why the focus of this project is to design possible OPV compounds to serve as the active layer in an OPV device.

In this study, compounds were designed as possible OPV compounds, and their highest occupied molecular orbital (HOMO) and lowest unoccupied molecular orbital (LUMO) values were determined to calculate their HOMO-LUMO gap (analogous to the band gap). Finding the HOMO-LUMO (HL) gap is an important parameter that helps determine the electrical properties of a material.[3] For example, the HL gap determines which wavelengths of light are absorbed and emitted by the material.[5]

The compounds designed are based on, naphthalene, 1,1'-biphenyl, and α-septithiophene as the backbones because they are some of the most efficient semi-conductors in the field of organic electronics.[3,6] As shown on Figure 7, EWGs such as 1,2,5-thiadiazole, 1,3,4-thiadiazole, thieno[3,4-c][1,2,5]thiadiazole (TTDZ), and other thienothiadiazole derivatives were attached to the polymer backbones. This was done because thiadiazoles are known to correlate to ideal energy level alignments for high-performance OPV compounds as they lower the HOMO-LUMO gap (HL gap).[7] Therefore, by combining those factors, five sets of compounds were designed: sets **a**, **b**, **c**, **d**, and **e** with variations within each set, for a total of 26 compounds. In addition, the calculated energy conversion efficiency versus the HL gap and the LUMO level was explored for the compounds that showed promising results.

Methods

Density Functional Theory (DFT) with the B3LYP functional and 6-311+G(d,p) basis set were used to calculate all the geometries at the ground state (S_0) level using the Gaussian 16 program on the Expanse supercomputer at the San Diego Supercomputer Center (SDSC).[8] The HOMO and LUMO values were then determined, and the HL gaps were calculated.

Results and Discussion

The compounds shown in Figure 1 are based on naphthalene as the semiconductor. In general, the HL gap decreases as more nitrogen atoms replace carbon atoms in the aromatic rings. Although the gap decreased significantly, it is still too high.

Set **b** compounds are shown in Figure 2, where 1,1'-biphenyl was used as the semiconductor backbone. Two biphenyl moieties were then attached at the 2 and 5 carbons of TTDZ. Nitrogen atoms then substituted several carbons in the biphenyl moieties to make 2,2-bipyrazine and other pyrazine and pyridine derivatives. Then, in compounds **b3-b6**, furan, and 1,3,4- oxadiazole were put in place of TTDZ. Replacing TTDZ with furan and 1,3,4-oxadiazole increased the HL gap significantly from 2.27 eV to 3.36 eV and 3.96 eV. Although it is uncertain, based on these results, it may be inferred that furan and its derivatives do not serve well in adjusting the HL gap in OPV compounds.

As seen in Figure 3, septithiophene was used as the semiconductor backbone. Benzene, pyrazine, and pyridine were conjugated to the thiophene rings of septithiophene. It was found that adding too many pyrazines to the semiconductor lowers the HL gap too much, whereas using pyridine lowers the HL gap to a favorable value. This can be seen in compound **c2** where pyrazine was used and the HL gap resulted in 0.98 eV, which is too low. Then, in **c3**, pyridine was used and resulted in a HL gap of 1.70 eV, which is a promising value. However, the LUMO value of **c3** was very low. Literature indicates a promising band gap is from 1.2 eV to 2.0 eV with a LUMO value of -3.6 eV to -4.0 eV.[9] Although, **c3** has a good HL gap, its LUMO value is insufficient with a value of -3.2 eV.

Set **d** compounds (Figure 4) were derived from compound **c3** because there are options for a lot of variabilities when using septithiophene as the semiconductor. Although pyrazine and pyridazine were also used in set **d** compounds, 3,4-dihydro-1,2,5-thiadiazole was conjugated to the thiophene rings of septithiophene to make a conjugated TTDZ system. It was found that having too many TTDZ's conjugated together lowers the HL gap too much. The average HL gap of the compounds in set **d** was 0.56 eV.

After evaluating the results of the HL gaps from set **a** compounds, where they were too high, and the HL gaps from set **d** were too low, set **e** compounds were designed strategically. As Figure 5 shows, septithiophene continued to be used as a semiconductor. However, less TTDZ was used, and a few benzene rings were placed with the intention to balance out the strong electron density pull from the TTDZ. It was found that two to four TTDZ moieties, conjugated with the semiconductor, resulted in proficient results. More specifically, compounds **e4**, **e5**, and **e6** came to have a HL gap of 1.1 eV to 1.5 eV and with a LUMO value of -3.7 eV to -3.9 eV. Both the HL gap and LUMO values were within ideal literature values.[9]

The compounds with HL gaps within acceptable parameters are plotted on the contour plot shown in Figure 6. This contour plot represents the bandgap (HL gap) on the x-axis and the LUMO value on the y-axis.[9] The contour lines demonstrate constant power conversion efficiencies. For example, compounds **e4** and **e6** would be about 8-9% efficient. Günes argues that a band gap (HL gap) of 1.1 eV can feasibly absorb 77% of solar photons, but do not mention the theoretical efficiency, whereas a band gap of 2 eV limits photon absorption to roughly 30%.[3] Regardless, these efficiencies are only theoretical, and there is a difference between efficiency and photon absorption.

Summary and Conclusions

An initial computational assessment was performed on possible OPV compounds. This initial computational assessment was based on comparing the HL gap to the LUMO gap. Various compounds were designed based on various semiconductors such as naphthalene, 1,1'-biphenyl, and α-septithiophene. The results discussed show that TTDZ is a highly effective EWG at lowering the HL gap. Based on the results, it can be concluded that compounds **b1**, **b2**, **c3**, **e1**, **e4**, **e5**, and **e6** showed the most efficient HL gap when compared to its LUMO value. However, of those seven compounds, only compounds **b2**, **e4**, **e5**, and **e6** will be pursued synthetically. Lastly, results indicate there is most likely a correlation between the amount of EWGs attached to a semiconductor and its calculated efficiency. This inference is stimulated by Figure 6, which has the seven most efficient compounds plotted. What can be analyzed from this plot is that compounds like **b1**, **c3**, and **e1**, that have minimal to no nitrogen atoms within the conjugated system and have few TTDZs attached, are the ones with the lowest efficiency. In comparison, the other four compounds that seem to have an appropriate ratio of TTDZ and nitrogen atoms are the ones with the highest efficiency. Further compounds ought to be designed and computed to further confirm this correlation.

Acknowledgments

Funding was received for this work from CSU-LSAMP SPaRa NSF grant number ***1826490.*** Computational time was provided by XSEDE (TG-CHE200068).

References

1. Kang, Y.; Youn, Y.; Han, S.; Park, J.; Oh, C.-S. Computational Screening of Indirect-Gap Semiconductors for Potential Photovoltaic Absorbers. *Chem. Mater.* **2019**, 31 (11), 4072–4080. https://doi.org/10.1021/acs.chemmater.9b00708.
2. Chen, L. X. Organic Solar Cells: Recent Progress and Challenges. *ACS Energy Lett.* **2019**, 4 (10), 2537–2539. https://doi.org/10.1021/acsenergylett.9b02071.
3. Günes, S.; Neugebauer, H.; Sariciftci, N. S. Conjugated Polymer-Based Organic Solar Cells. *Chem. Rev.* **2007**, 107(4), 1324–1338. https://doi.org/10.1021/cr050149z.

4. Latunussa, C. E. L.; Ardente, F.; Blengini, G. A.; Mancini, L. Life Cycle Assessment of an Innovative Recycling Process for Crystalline Silicon Photovoltaic Panels. *Solar Energy Materials and Solar Cells* **2016**, 156, 101–111. https://doi.org/10.1016/j.solmat.2016.03.020.
5. Perzon, E.; Wang, X.; Admassie, S.; Inganäs, O.; Andersson, M. R. An Alternating Low Band-Gap Polyfluorene for Optoelectronic Devices. *Polymer* **2006**, 47 (12), 4261–4268. https://doi.org/10.1016/j.polymer.2006.03.110.
6. Coropceanu, V.; Cornil, J.; da Silva Filho, D. A.; Olivier, Y.; Silbey, R.; Brédas, J.-L. Charge Transport in Organic Semiconductors. *Chem. Rev.* **2007**, 107 (4), 926–952. https://doi.org/10.1021/cr050140x.
7. Hachmann, Aspuru-Guzik et al. Lead candidates for high-performance organic photovoltaics from high-throughput quantum chemistry – the Harvard Clean Energy Project. *Energy Environ. Sci.* **2014**, 7 (2), 469. DOI: 10.1039/c3ee42756k
8. Gaussian 16, Revision C.01, M. J. Frisch, G. W. Trucks, H. B. Schlegel, G. E. Scuseria, M. A. Robb, J. R. Cheeseman, G. Scalmani, V. Barone, G. A. Petersson, H. Nakatsuji, X. Li, M. Caricato, A. V. Marenich, J. Bloino, B. G. Janesko, R. Gomperts, B. Mennucci, H. P. Hratchian, J. V. Ortiz, A. F. Izmaylov, J. L. Sonnenberg, D. Williams-Young, F. Ding, F. Lipparini, F. Egidi, J. Goings, B. Peng, A. Petrone, T. Henderson, D. Ranasinghe, V. G. Zakrzewski, J. Gao, N. Rega, G. Zheng, W. Liang, M. Hada, M. Ehara, K. Toyota, R. Fukuda, J. Hasegawa, M. Ishida, T. Nakajima, Y. Honda, O. Kitao, H. Nakai, T. Vreven, K. Throssell, J. A. Montgomery, Jr., J. E. Peralta, F. Ogliaro, M. J. Bearpark, J. J. Heyd, E. N. Brothers, K. N. Kudin, V. N. Staroverov, T. A. Keith, R. Kobayashi, J. Normand, K. Raghavachari, A. P. Rendell, J. C. Burant, S. S. Iyengar, J. Tomasi, M. Cossi, J. M. Millam, M. Klene, C. Adamo, R. Cammi, J. W. Ochterski, R. L. Martin, K. Morokuma, O. Farkas, J. B. Foresman, and D. J. Fox, Gaussian, Inc., Wallingford CT, **2019**.
9. Scharber, M. C.; Mühlbacher, D.; Koppe, M.; Denk, P.; Waldauf, C.; Heeger, A. J.; Brabec, C. J. Design Rules for Donors in Bulk-Heterojunction Solar Cells—Towards 10 % Energy-Conversion Efficiency. *Advanced Materials* **2006**, 18 (6), 789–794. https://doi.org/10.1002/adma.200501717 .

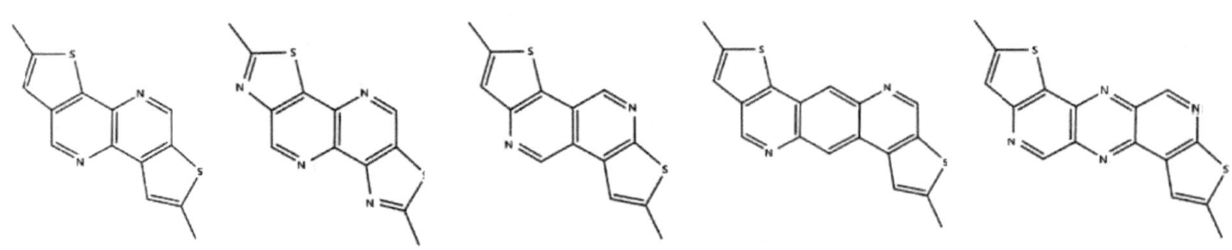

a1:	**a2:**	**a3:**	**a4:**	**a5:**
4.211376 eV	**4.460800 eV**	**3.963856 eV**	**3.572720 eV**	**3.065168 eV**

Figure 1. Set a compounds and their HOMO-LUMO gap computed at B3LYP/6-311+G(d,p).

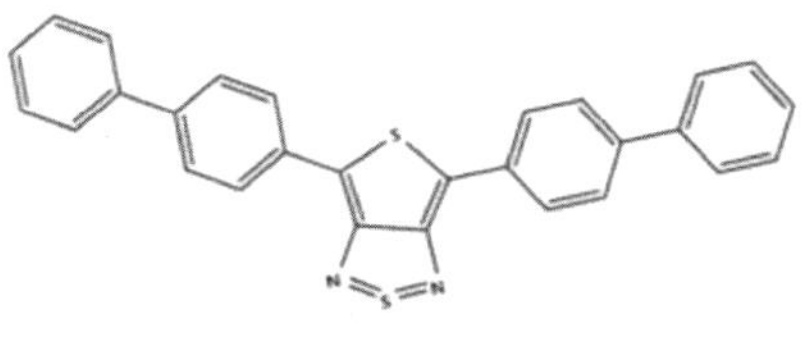

b1:

2.1 eV

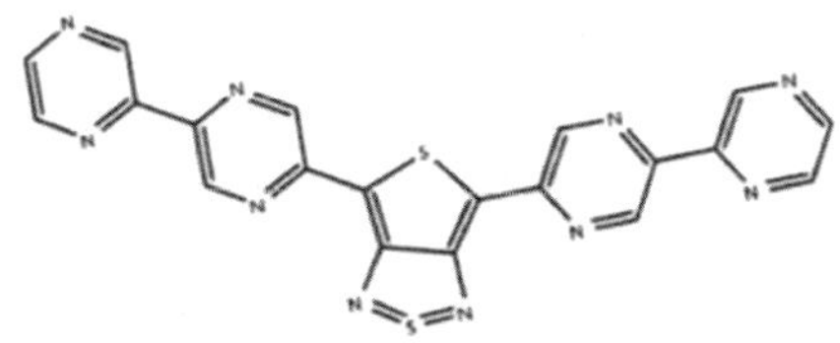

b2:

2.2 eV

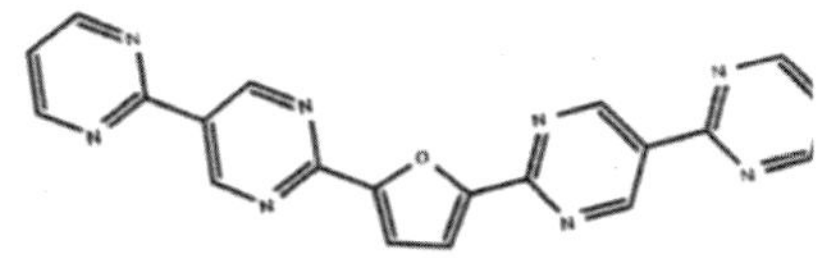

b3:

3.3 eV

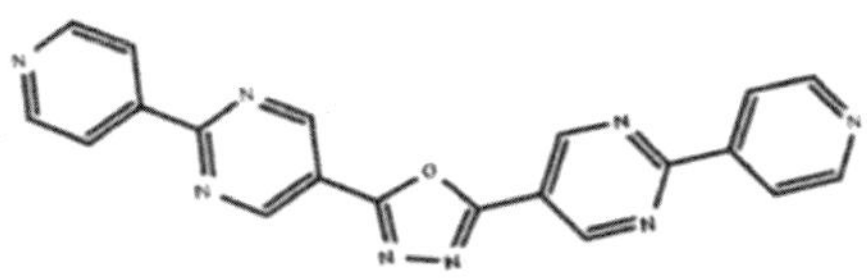

b4:

3.9 eV

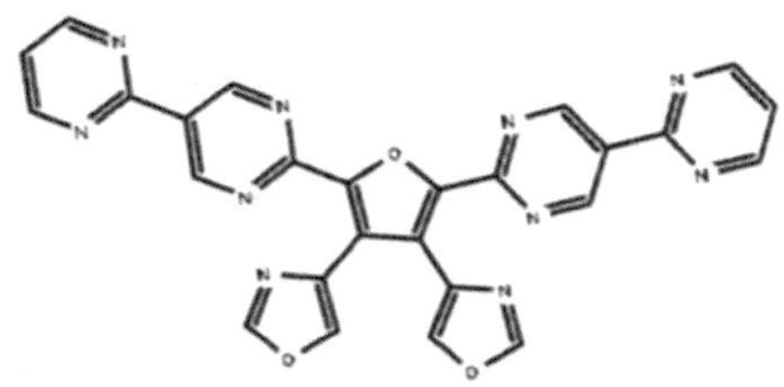

b5:

3.3 eV

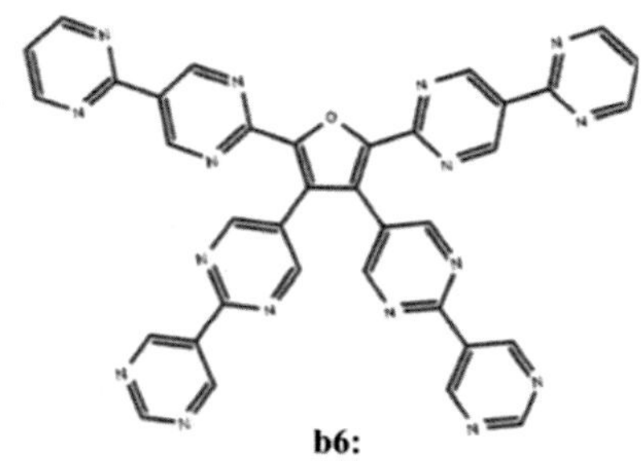

b6:

3.4 eV

Figure 2. Set b compounds and their HOMO-LUMO gap computed at B3LYP/6-311+G(d,p).

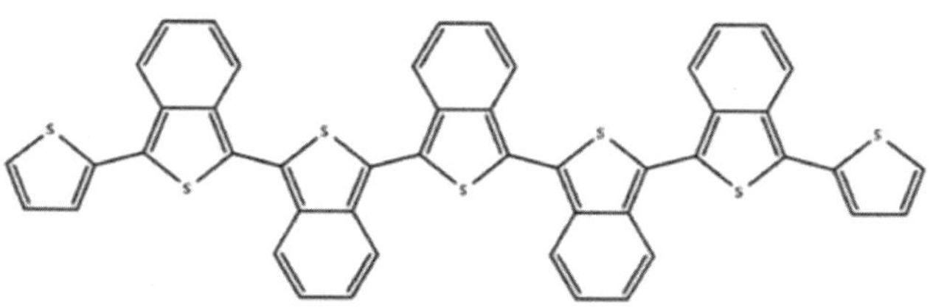

c1:

2.2 eV

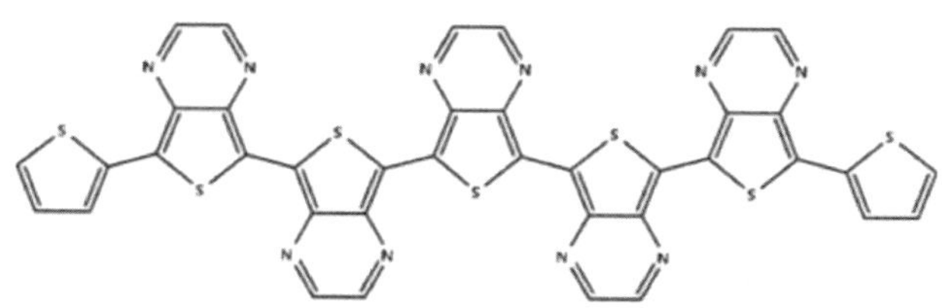

c2:

0.98 eV

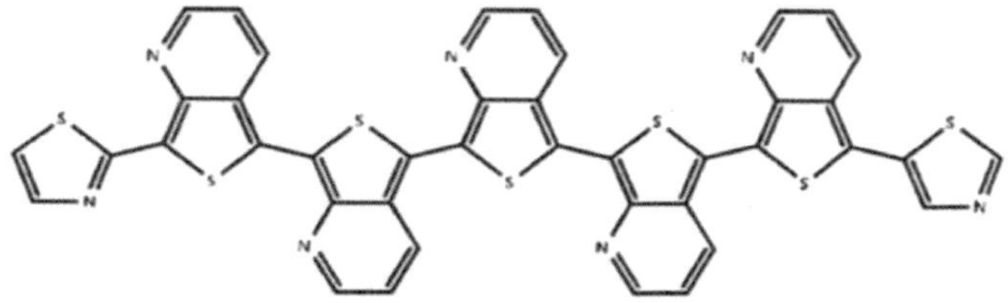

c3:

1.7 eV

Figure 3. Set c compounds and their HOMO-LUMO gap computed at B3LYP/6-311+G(d,p).

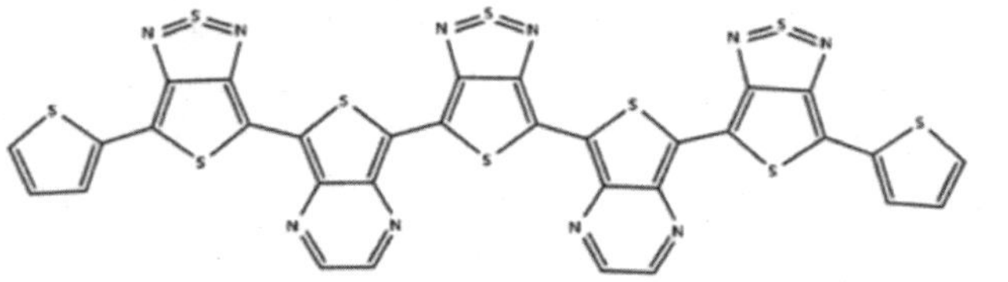

d1: 0.46 eV

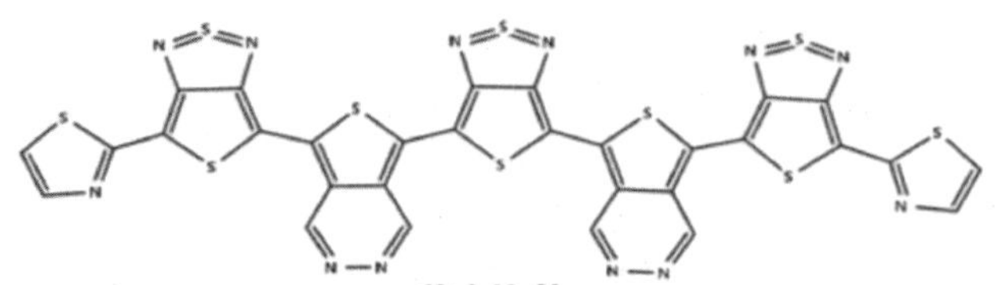

d2: 0.66 eV

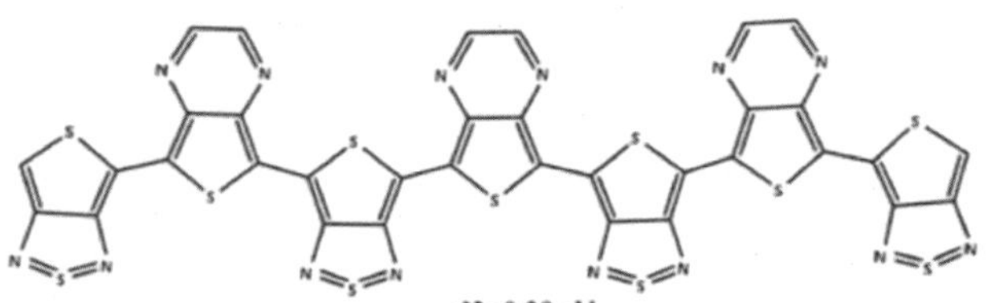

d3: 0.39 eV

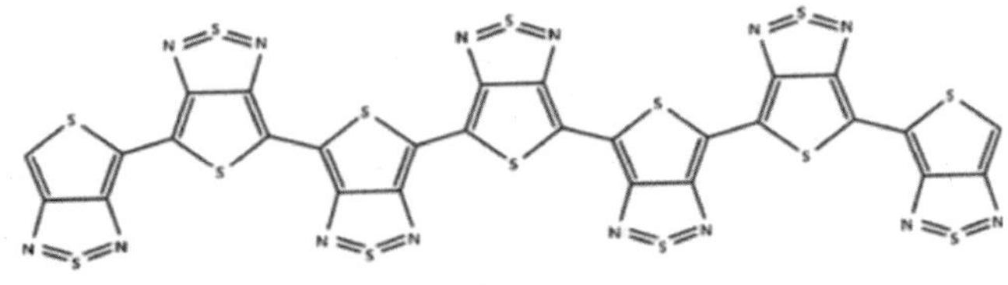

d4: 0.21 eV

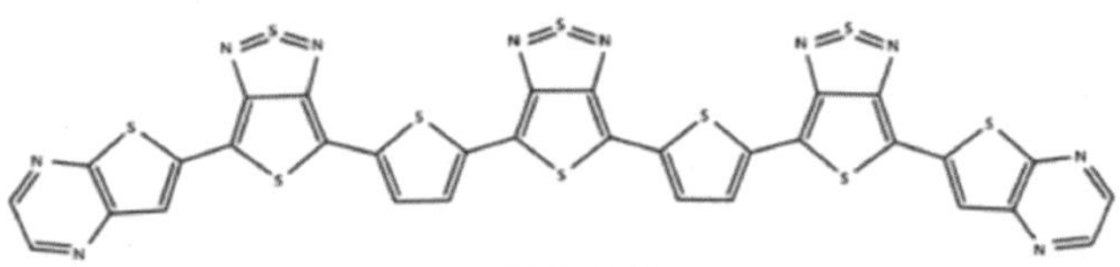

d5: 0.95 eV

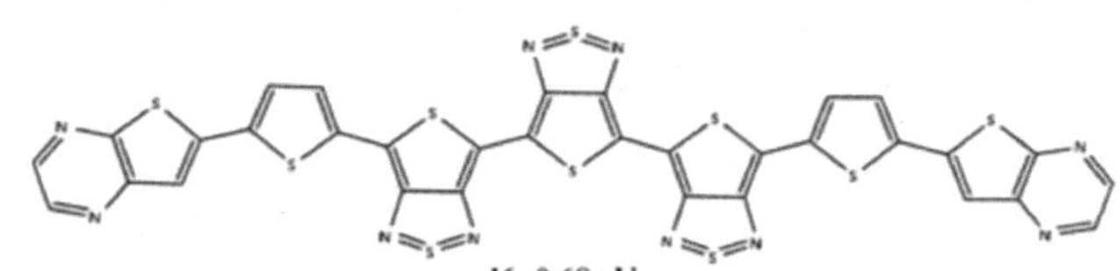

d6: 0.68 eV

Figure 4. Set d compounds and their HOMO-LUMO gap computed at B3LYP/6-311+G(d,p).

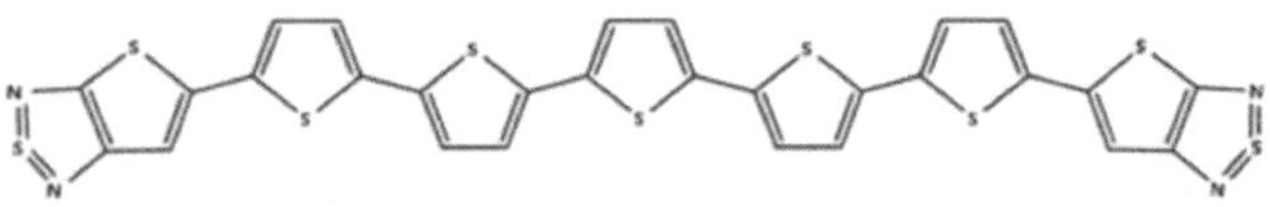

e1: 2.3 eV

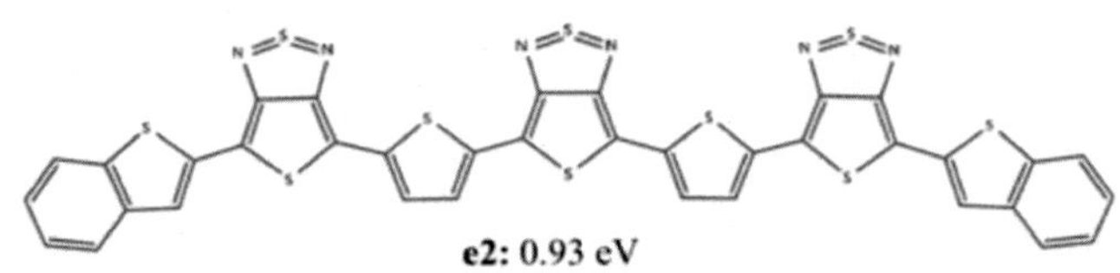

e2: 0.93 eV

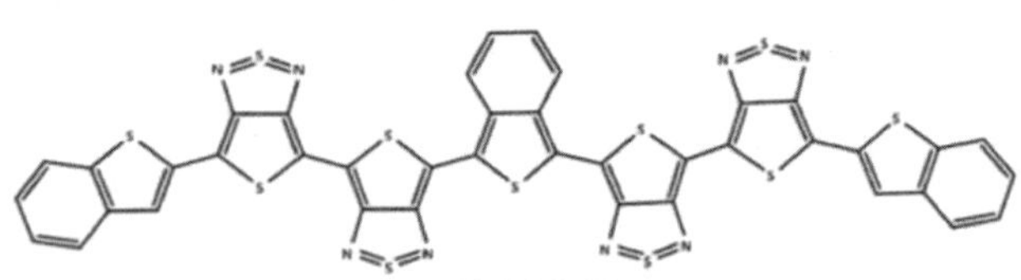

e3: 0.36 eV

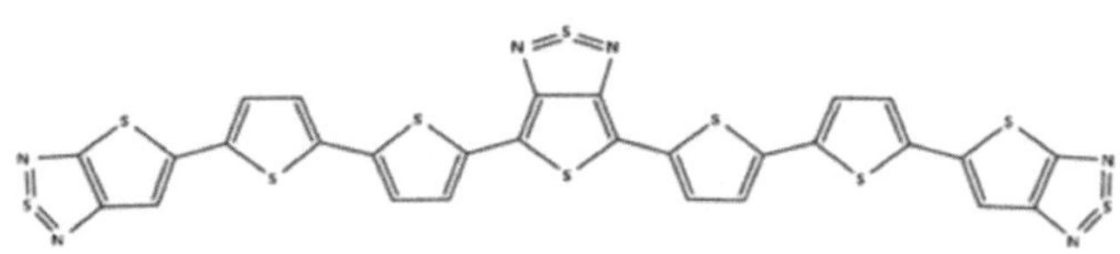

e4: 1.4 eV

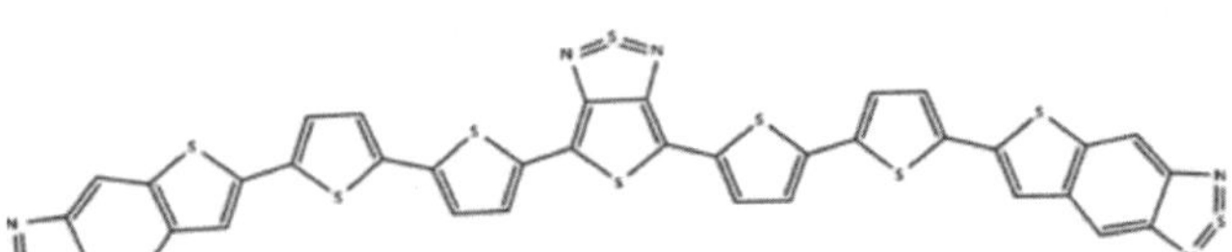

e5: 1.4 eV

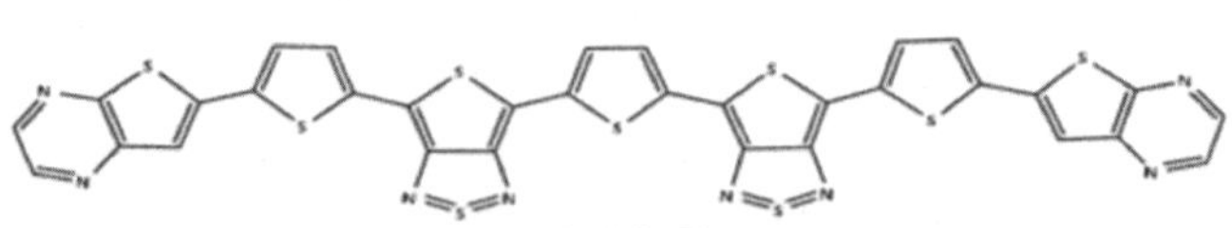

e6: 1.1 eV

Figure 5. Set e compounds and their HOMO-LUMO gap computed at B3LYP/6-311+G(d,p).

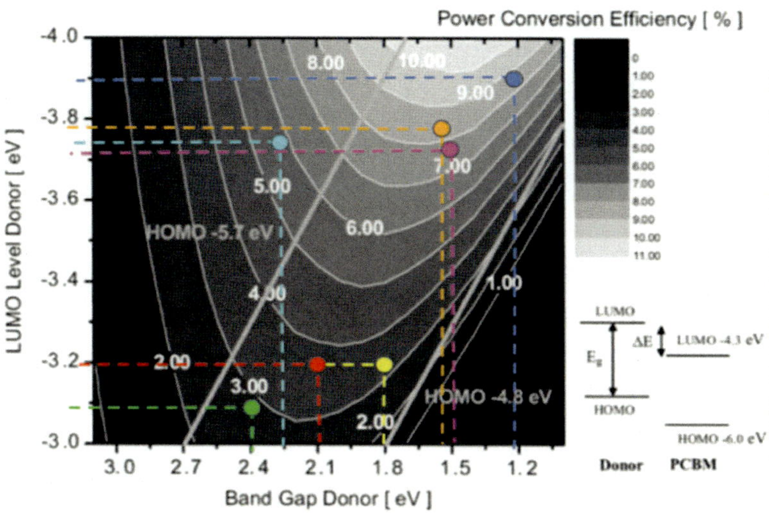

Compound #	LUMO (eV)	HOMO-LUMO Gap	Color on contour plot
b1	-3.2	2.1	red
b2	-3.7	2.2	light blue
e3	-3.2	1.7	yellow
e1	-3.2	2.3	green
e4	-3.7	1.4	orange
e5	-3.7	1.4	purple
e6	-3.9	1.1	dark blue

Figure 6. Estimated power conversion efficiency of seven of the compounds designed. Adapted from Scharber and collaborators.[5]

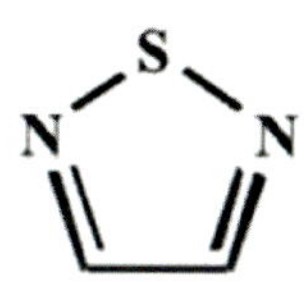

1,2,5-thiadiazole

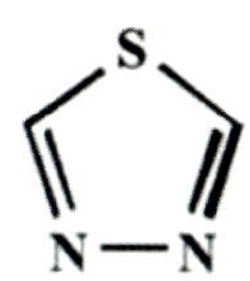

1,3,4-thiadiazole

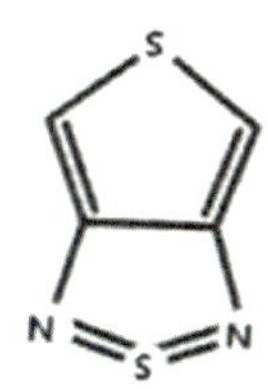

thieno[3,4-c] [1,2,5]thiadiazole

Figure 7. EWG's used.

Injustice Within Renewable Energy Life Cycles: Can IRENA Offer a Solution?

G. Webster Ross

Abstract

This paper explores case studies of green colonialism, supply chain injustices, and poor e-waste management within renewable energy life cycles, and investigates how the International Renewable Energy Agency (IRENA) may be the best suited organization to address and mitigate these issues on a global scale. While renewable energy technology is often heralded as the key to a sustainable future, the life cycle of these technologies is riddled with human rights violations and other injustices. To begin with, many of the minerals required for assembling the hardware are mined in unregulated environments, resulting in several injustices such as health and safety hazards for miners, child labor, and insufficient pay and protection for the miners. Furthermore, many manufacturing factories for renewable energy hardware have unregulated emissions, creating a hazardous environment for communities living near the factories. During the construction phase of the renewable energy life cycle, there are many cases of the land required for the projects being stolen from Indigenous populations through force, coercion, or political maneuvering, thus putting more unnecessary burdens on communities who have faced centuries of oppression and marginalization. Finally, at the end of life of the renewable energy tech, the hardware is sent to e-waste scrapyards in low-income countries where human rights violations similar to those seen in the mining industry are commonplace. Over the first decade of its existence, IRENA has had unprecedented success in creating an international community supporting knowledge sharing of renewable energy policy and construction best practices. As a result, it has the collaborative infrastructure and information pathways required to quickly brainstorm and disseminate policies to manage and mitigate these poignant issues surrounding renewable energy. By increasing focus on energy justice, pursuing active collaboration with Indigenous Nations, and encouraging reduced energy consumption in Western countries, IRENA could become a key leader in a globalized energy justice movement that would not only save countless lives and livelihoods, but also help to legitimize renewable energy's promise of a sustainable future.

Keywords: Energy justice, Green colonialism, E-waste, IRENA, Renewable energy, Wind energy

Introduction

While renewable energy technologies hold much promise for reducing global emissions and thus mitigating the harmful effects of climate change, the manufacturing, construction, and decommissioning of renewable development and technology have harmful externalities that have yet to be properly addressed. Moreover, the burdens of these negative externalities, similar to those perpetuated by the fossil fuel industry, are disproportionally felt by communities that are predominantly black, Indigenous, and people of color (BIPOC). From deplorable conditions at mining sites in the Democratic Republic of the Congo (Sovacool et al 2020) to the continued dispossession of Indigenous lands and resources in the name of green energy (Normann 2020; Dunlap 2018), the renewable energy industry is following in the footsteps of the fossil fuel industry by exploiting and oppressing BIPOC communities in the name of progress. These issues are inextricably tied to the global transition towards renewable energy and subsequently need to be acknowledged and addressed on a global scale. As the leading international advocate and policy advisor for renewable energy projects, the International Renewable Energy Agency (IRENA) has the proper foundation to build an international energy justice platform and spearhead efforts to mitigate the harmful and disproportionally felt impacts of the renewable energy industry on a global scale. In this paper, after outlining a few of the major justice issues within the renewable energy industry, I offer three policy recommendations for IRENA's approach to energy justice that would support a more sustainable approach.

Background

Definition of Key Terms

Throughout this paper, I bring up the term green colonialism to describe the effect some renewable energy projects have on Indigenous nations. Green colonialism is a subset of settler colonialism, and to fully understand the former, it is important to adequately comprehend the latter. *Settler colonialism* differs from other forms of colonialism, it is "a structure designed to eliminate the Native via physical and political erasure" (Gilio-Whitaker 2019, 12). Whereas other colonizing models seek control over the Indigenous peoples, lands, and resources, settler colonizers "seek to inscribe their own homelands over Indigenous homelands, thereby erasing the history, lived experiences, social reality and possibilities of a future of Indigenous peoples" (Whyte 2016). Erasing entire nations requires coordination and systematic violence; Indigenous scholar and activist Dina Gilio-Whitaker poignantly points out that "settler colonialism, with its mandate to eliminate the Native, is fundamentally genocidal" (Gilio-Whitaker 2019, 50). However, Indigenous peoples have proven to be resilient in their efforts of survival, despite the fact that settler colonialism is not just a moment in history but "centuries of genocidal policies, treaty violations, illegal land seizures, and environmental catastrophes perpetuated by the [...] settler government" (Gilio-Whitaker 2019, 5). Ojibwe scholar and activist Winona LaDuke explains that these oppressive strategies are ultimately "for the purpose of 'developing' the [settler] economies and, subsequently, the 'underdeveloping' of Indigenous communities" (LaDuke 1994, 131). I later explain how green colonialism ties into this structure of settler colonialism.

Another concept I discuss is the *sustainability* of certain systems and projects. It is important to distinguish between the terms "renewable" and "sustainable." Renewable energy is characterized by the fact that the source of energy is un-consumable. Sustainable energy, on the other hand, is energy generation that could be continued indefinitely with little to no impact on the surrounding ecological bodies. For example, hydroelectric dams are often considered renewable energy. However, they are not sustainable, as dams can impact immigration patterns of river life, flood important habitats upstream of the dam, and endanger habitats downstream of the dam by reducing the water supply. A more complete interpretation of sustainability can be drawn from Indigenous teachings and practices or Traditional Ecological Knowledge (TEK). In her article "Traditional ecological knowledge and environmental futures," scholar-activist Winona LaDuke introduces a concept central to the Anishinaabe lifestyle:

> "'Minobimaatisiiwin,' or the 'good life,' is the basic objective of the Anishinabeg and Cree people who have historically, and to this day, occupied a great portion of the north-central region of the North American continent. [...] This is how we traditionally understand the world and how indigenous societies have come to live within natural law. Two tenets are essential to this paradigm: cyclical thinking and reciprocal relations and responsibilities to the Earth and creation. Cyclical thinking [...] is an under-

> standing that the world [...] flows in cycles. Within this understanding there is a clear sense of birth and rebirth and a knowledge that what one does today will affect one in the future, on the return. A second concept, reciprocal relations, defines responsibilities and ways of relating between humans and the ecosystem. Simply stated, the resources of the economic system [...] are recognized as animate and, as such, gifts from the Creator. Within that context, one could not take life without a reciprocal offering..." (LaDuke 1994, 128).

The belief in a reciprocal responsibility between humans, non-humans, and the land is not unique to the Anishinaabe people. In her book *Aloha Betrayed,* Kanaka scholar Noenoe Silva describes the Hawaiian concept of "pono," explaining that "in the ancient Kanaka world, pono meant that the akua, (deities) ali'i, kahuna, maka'āinana, and 'āina (land) lived in balance with each other and that people had enough to eat and were healthy" (Silva 2004, 16). The concepts of Minobimaatisiiwin and pono affirm the statement made by Indigenous scholar Dina Gilio-Whitaker, that "the Indigenous world is a world of relationships built on reciprocity, respect, and responsibility, not just between humans but also extending to the entire natural world" (Gilio-Whitaker 2019, 138). This belief system is what I base my translation of sustainability on. To be considered sustainable, a practice or technology must be balanced and equitable throughout all stages of its life cycle for all humans, non-humans, and resources involved. For more in-depth information on TEK and Indigenous perspectives, I encourage readers to explore works by other Indigenous scholars such as Noelani Goodyear-Ka'ōpua, Deborah McGregor, and John Borrows.

Fossil Fuels vs. Renewable Energy

Fossil fuels are characterized by their ability to create energy through combustion. When fossil fuels are burned, in addition to providing energy, they release carbon dioxide and other gases into the atmosphere. Since the mid-20th century, scientists have been raising the question of whether the gases released by burning fossil fuels, referred to as greenhouse gases, contribute to the rapid global warming Earth has been experiencing over the past century. By the 21st century, there was a strong scientific consensus on anthropogenic global warming (AGW), with a 2013 study published in *Environmental Research Letters* stipulating that of nearly 12,000 articles in peer-reviewed scientific literature from 1991 to 2011 that expressed a position on AGW, "97.1% endorsed the consensus position that humans are causing global warming" (Cook et al. 2013). A similar study done in 2021, which analyzed 88,125 peer-reviewed articles from 2012 to the present, concluded that "the scientific consensus on human-caused contemporary climate change—expressed as a proportion of the total publications—exceeds 99% in the peer reviewed scientific literature" (Lynas et al. 2021). With an unprecedentedly strong scientific consensus, it is fair to say that burning fossil fuels does contribute to global warming, and if the global consumption of fossil fuels is not reduced, then the world faces an exponentially intensifying ecological crisis.

Global warming is not the only harmful externality of fossil fuels. The extraction of fossil fuels like coal, oil, and natural gas is often an invasive process that has detrimental effects on the surrounding environment and communities. For example, a common method of coal extraction in the eastern United States is mountaintop removal (MTR), which not only destroys the habitat of countless animal species but also contaminates surrounding rivers and streams that are the main water source for local communities (Kaneva 2010; Boyles et al. 2017). Additionally, the discovery of fossil fuel deposits suddenly made those lands highly coveted, which resulted in another wave of land dispossession from Indigenous peoples in settler states such as the United States and Canada (Fixico 2021; Huseman & Short 2012; Preston 2017). Finally, power plants fueled by fossil fuels have a long history of emitting toxic gases into the atmosphere and dumping toxic byproducts into nearby bodies of water, thereby making living conditions for local communities deplorable (Bullard 1994; Bullard 2005; Pulido et al. 1996). To top it off, these harmful impacts are disproportionally felt by communities who are black, Indigenous, and people of color (BIPOC), so much so that the term "environmental racism" was coined to describe such inequities (Bullard 1994; Bullard 2005; Holifield 2001; Pulido et al. 1996). These issues are just as important as greenhouse gas emissions, and any transition away from fossil fuels should also be a transition away from unsustainable resource extraction, land dispossession, and an unequal share of burdens.

Renewable energy, as mentioned before, is characterized by the process of creating energy through non-consumable natural resources such as the sun or wind. Since renewable energy doesn't involve burning any consumable fuel, there are no emissions involved with capturing the energy. It should be noted that while there are no emissions related to renewable energy generation, there are emissions associated with the life cycle of renewable energy due to resource extraction, manufacturing of the technology, and the transportation/construction of infrastructure. Some examples of renewable energy technology are wind turbines, solar panels, and geothermal plants. As a consensus on anthropogenic global warming (AGW) has grown over the past few decades, resources have poured into renewable energy research and development, making the technology more advanced and affordable. For example, the average price of solar panels has gone from $4.90 per watt of capacity in 2000 to just $0.20 per watt of capacity by 2020 (IEA 2020). As a result, installed renewable capacity has exploded over the past twenty years, with global solar energy generation going from 1.1 TeraWatts (TW) in the year 2000 to 855.7 TW in the year 2020; and global wind energy generation rising from 31.4 TW in 2000 to 1591.2 TW in 2020 (BP 2020). In the face of global warming, this is a promising trend that should be maintained. In an effort to expedite a global transition to renewable energy, several institutions and organizations have formed initiatives around encouraging and facilitating further installation of renewable energy technology. While most of these initiatives have been subsets of larger bodies such as a sub-department within the International Energy Agency, there has been one international organization created for the specific purpose of supporting renewable energy growth: the International Renewable Energy Agency.

The International Renewable Energy Agency

Created in 2009, the International Renewable Energy Agency (IRENA) made history by becoming "the first intergovernmental organization exclusively focused on renewable energy" (Overland 2018, 336). Starting with 75 initial signatories, the organization's relevance grew rapidly, and by 2013, the members and applicants for membership in IRENA amounted to a total of 161 states (Urpelainen 2015; Mengi-Dincer 2021). The organization is comprised of three main bodies: the Assembly, which is the "ultimate decision-making authority, made up of one representative from each Member State"; the Council, a group "of 21 Member States elected for a two-year term"; and the Secretariat, which "provides administrative and technical support to the Assembly, the Council, and their subsidiary bodies" (IRENA 2021). IRENA's main focus is to serve as a "principal platform for international co-operation, a center of excellence, and a repository of policy, technology, resource and financial knowledge on renewable energy" (IRENA 2021). Since the organization's focus isn't directly funding or implementing projects, it is referred to as an "epistemic" organization or one that deals mostly with knowledge-sharing and collaboration between member states (Urpelainen 2015).

While IRENA is still a relatively young organization, scholars who have written on IRENA generally agree that it has had unprecedented success in its early stages and promises to grow in relevance and impact over the coming years. In their 2015 article in the journal *International Environmental Agreements*, professors Johannes Urpelainen and Thijs Van de Graaf make an early evaluation of IRENA's role in global energy governance. They define IRENA as an epistemic organization and list some of IRENA's main activities, such as "lowering informational barriers and asymmetries, gathering and disseminating knowledge, and comparing and evaluating national regulatory frameworks to identify best practices in renewable energy governance" (Urpelainen 2015, 168). Ultimately, Urpelainen and Van de Graaf conclude that since "IRENA has, in spite of a small budget and the lack of a proven track record, established itself as a major provider of epistemic services to the least developed countries" (Urpelainen 2015, 174), the organization "can be regarded as a success story in institutional innovation" (Urpelainen 2015, 161). Scholars Indra Overland and Gunilla Reischl provide a more recent evaluation of IRENA's role in global energy governance, measuring IRENA's success using the following three questions: What types of representatives do member states send to IRENA meetings, and what does this reveal about how IRENA is seen as an organization? What financing and Human Resources does IRENA have access to? And how often is IRENA mentioned in national energy policy documents? (Overland 2018). Overland and Reischl come to a similar conclusion that IRENA has had impressive early success, noting that in terms of mentions in international energy policy documents, "IRENA has rapidly caught up with the IEA (International Energy Agency) in the renewable energy niche, achieving parity in 2014" (Overland 2018, 348). They also predicted that "renewable energy seems set to grow in importance, and along with it, so will IRENA" (Overland 2018, 348). Finally, scholar Mengi-Dincer, along with professors Volkan S. Ediger and Cagla Gul Yesevi, evaluate IRENA through the lens of social constructivism, focusing on the various norms that IRENA is setting within the global energy governance field (Mengi-Dincer 2021). They too do not hold back on emphasizing IRENA's importance within global governance, concluding that "IRENA is found to play an important role in developing renewable energy policies worldwide as well as in encouraging its members to embrace a new paradigm for their energy preferences by creating knowledge, shaping behaviors, and changing norms in the global energy system" (Mengi-Dincer 2021, 8).

Justice Issues Within a Renewable Energy Transition

While renewable energy is vital for reducing the impact of global warming, we need to look at the sustainability of these technologies with a critical eye to avoid falling into the same harmful and oppressive patterns that were normalized by the fossil fuel industry. It is easy to rationalize the negative impacts of renewable energy by referencing the urgency of the ecological crisis (and thus the importance of maximizing renewable installations at all costs), but it is essential to recognize that it was that type of thinking that created the ecological crisis to begin with. Greenhouse gases may have been the catalyst that started rapid global warming, but the original cause of our ecological crisis was prioritizing costs over sustainability, rashly acting without thinking of future costs, commoditizing land and resources, and de-mattering BIPOC lives and cultures. Blindly building as many renewable energy projects as possible without considering whether they are done the right way will only set the world up for a different crisis down the road. Therefore, it is imperative that the social justice issues of the fossil fuel industry are left in the past. Unfortunately, the following section shows that, in some ways, the renewable energy industry is still perpetuating these toxic norms.

Green Colonialism

Green colonialism is a recently coined term used to describe how the renewable energy industry, just like the fossil fuel industry, is "intensify[ing] colonial losses of land and rights" of Indigenous populations around the world (Normann 2020, 78). Before getting into the case studies of green colonialism, I feel it is important to explain why green colonialism is wrong. A possible rationalization for green colonialism is that while further land dispossession may not be ideal, it is for "the greater good" of humanity. Professor and scholar Bruce Erickson points out the flaws in this theory by analyzing the title scholars have given our current geological epoch: the Anthropocene. As mentioned in the background section of this article, it is widely accepted that rapid climate change is a direct result of human activity over the past two centuries; hence the geological epoch we are experiencing due to climate change is commonly referred to as "the Anthropocene." As Erickson explains, the term Anthropocene looks to the past and the future: declaring humans as both the cause of the environmental crisis and its solution(Erickson 2018). Erickson also points out that "the Anthropocene [...] is dependent upon a universal image of the Anthropos," that is, a white, Euro-centric Anthropos (Erickson 2018, 3). The issue with a universal image of the Anthropos is that it conveniently paints over the existence of non-white humans and depicts the entire human population as a unified entity *equally responsible* for the environmental crisis, which we know to be untrue since the vast majority of fossil fuels were extracted and burned for the sole benefit of white economies such as those of the United States, Canada, and Europe. This is important because the homogenization of humanity and their responsibility for global warming implies that we must all be unified in the strategy for how to go about reversing the effects of the crisis: a strategy that is conveniently designed and implemented for the most part by white governments. This "for the greater good" argument acts as a rationalization for settler governments to "circumscribe Indigenous agency in the interest of a greener future," thus positioning the settler government to further "assert jurisdiction over [the] land, and not just claim it, [which] lies at the heart of the structure of settler colonialism" (Erickson 2018, 4). In short, Indigenous people should not be sacrificed for a solution to a problem that they hardly contributed to in the first place, and no amount of renewable energy capacity will justify the continued oppression and violence against Indigenous peoples. To illustrate my point, I refer to the following two case studies: the Southern Saami tribe in Norway and the Indigenous tribes within the Isthmus of Tehuantepec, Oaxaca, in southern Mexico.

The Saami are an Indigenous people whose ancestral territory spans across Sweden, Norway, Russia, and Finland (Normann 2020, 80). There are several sub-tribes within the Saami, one of which is the Southern Saami, who live across Norway and Sweden. The Southern Saami currently consists of an estimated "population of around 2000 people [...] [which] includes approximately 500 native language speakers," and reindeer herding lies at the heart of their cultural heritage (Normann 2020, 80). Over the years, "colonial and state assimilation practices have affected their community, leaving them with few remaining spaces to strengthen and transfer knowledge, language, and cultural practices, except those generated around herding" (Normann 2020, 80), which has put an even higher importance on the few pasturelands left available for the herders to keep their reindeer. Unfortunately, these mountainous lands have also caught the eye of wind energy developers, and subsequently, "Southern Saami lands have additionally turned into sites of contestation over wind energy development" (Normann 2020, 80). While the Saami insist that building the wind turbines "bring increased human activity, the construction of energy infrastructure, and new road networks that will negatively affect reindeers' pasturelands" (Normann 2020, 81), lawyers representing the renewable energy projects' interests contest "whether or not the wind turbines will deter herding [...] thus downscaling the value of Saami knowledge" (Normann 2020, 91). The dismissive air of governments and corporations is not new to the Saami people, but that doesn't make it any less frustrating or painful. In nearly all of the interviews Normann had with Saami people, words like "cultural genocide" and "racism" were used to describe the land dispossession caused by wind turbine projects (Normann 2020, 89). These accusations should not be taken lightly, and if renewable energy projects are pushing cultures on the brink of extinction, then there is nothing sustainable about them.

The situation in the Isthmus of Tehuantepec is no better. The Isthmus has been heralded as having 'the best wind resources on earth' (IFC 2014) and as a result, has seen a huge influx of investors and contractors itching to exploit this suddenly valuable resource. The projects first started in Zapotec territory, the northern region, gaining support from the local and Indigenous communities with promises of "work, social development, and prosperity" (Dunlap 2018, 558). Unfortunately, after the projects were completed, "many of these promises remained unfulfilled, limited and benefited a minority of the populations" (Dunlap 2020, 558). Furthermore, unanticipated negative consequences began to arise from the land-use change, including "altered agricultural and livestock patterns, [...] the clearing of animal habitat, compacting of soil for roads, loss of birds, transforming the ground water into concrete for wind turbine foundations, and, finally, leaking oil into the ground, which people claimed contaminated both the ground water and animals" (Dunlap 2018, 559). As the wind sites spread to Ikoot territory in the south, Indigenous communities started putting up more resistance, but to avoid slowing construction, "public consultation was bypassed, instead opting [...] for selective negotiations with select regional administrators, elites and social property members" (Dunlap 2018, 559). To make matters worse, projects proposed building windmills on the coast and within the ocean, which is the main source of subsistence for the Ikoot people. During one pilot wind turbine installation, witnesses reported that the first attempt at building a foundation for the turbine "resulted in the mass killing of fish as far as the eye could see" (Dunlap 2018, 560). During his interviews with the Zapotec and Ikoot people, Dunlap also noticed the common comparison of the wind projects to ethnocide and genocide (Dunlap 2018, 550). Similar to the people of the Southern Saami tribe, for the Ikoot and Zapotec people, the combat against the construction of more wind turbines is more than just resistance; it "is conceived as a war devised to 'annihilate' them, which is seen as a generational fight" (Dunlap 2018, 564). For too long, the voices of Indigenous communities like the Southern Saami, Zapotec, and Ikoot have been ignored or silenced through political maneuvering, gaslighting, and violence. For renewable energy projects to be sustainable, Indigenous voices need to be elevated on an international scale, and Indigenous representatives brought into the planning and collaboration circles to protect the rights and well-being of Indigenous lands, resources, and people.

Upstream & Downstream Justice Issues

Besides acting as a potential tool of green colonialism, renewable energy technology itself has many justice and equity issues embedded within its upstream (mineral extraction and manufacturing) and downstream (waste management) processes. Sovacool et al. describe the injustices within these practices as the 'decarbonisation divide' since the benefits and negative externalities associated with the system are imbalanced between Global North and Global South countries (Sovacool et al. 2020, 2). Specifically, Sovacool et al. focus on cobalt mining in the Democratic Republic of the Congo (DRC) since cobalt is a key component of many renewable energy technologies, including wind turbines, battery storage, and e-waste management within Ghana. Upon detailed analysis of mineral extraction in the DRC and e-waste management in Ghana, the researchers identified "environmental and public health risks; gender discrimination and the marginalization of women; child labor and exploitation; and the subjugation of ethnic groups" (Sovacool et al. 2020, 7) as the primary issues. In the cobalt mines, it is common to see children working "underground, underwater, at dangerous heights, or in confined spaces [...] routinely carry[ing] sacks of ore that weigh more than they do" (Sovacool et al. 2020, 14). The mines are unregulated and have no safety precautions, and "many [children] will die before then ever become an adult. They will get buried alive in an underground tunnel, or drowned in a waterlogged pit. [...] they can even develop cancer, things like pneumonia, malnutrition, or they start dying from AIDS" (Sovacool et al. 2020, 14). As if the conditions of the mines were not bad enough, "children are also often exposed to physical abuse and bearings, whippings, and attempted drownings from security guards, as well as drug abuse, violence, and sexual exploitation" (Sovacool et al. 2020, 14). Conditions in the e-waste dumpsite in Ghana are no better. Locals interviewed by the researchers report seeing "children sleeping on scrap, eating with e-waste, coughing intensely, [and] bleeding" (Sovacool et al. 2020, 14). Children as young as nine years old pick through the waste and burn it with no protective gear, exposing themselves to toxic chemicals and noxious fumes that contribute to "abnormally high rates of spontaneous abortions, stillbirths, and premature births" in the area (Sovacool et al. 2020, 13). While renewable energy technology does not make up the entirety of e-waste, it promises to exponentially contribute to the global e-waste inventory, with waste from end-of-life solar products alone projected to reach "a worldwide total of 60 to 78 million tons of waste" by 2050 (Sovacool et al. 2020, 4). To put that value in perspective, that amount of waste "would make solar PV waste flows greater than *all* e-waste flows in 2018" (Sovacool 2020, 4).

In addition to injustices and inequalities within mineral extraction and waste management, many renewable energy technologies such as solar panels or wind turbine parts "are manufactured with no environmental or public health regulation in poor Global South communities, exposing the people of color who work solar factory assembly lines and live in factory-adjacent homes to a host of deleterious toxins and pollutants that severely compromise their health and well-being" (Lennon 2018, 23). These issues of exploitation, emissions, and waste within renewable energy technology processes will only worsen as more renewable energy capacity is built. IRENA has estimated that in order to keep global warming within safe levels, "the number of electric vehicles (EVs) needs to jump from almost one million in 2015 to one billion cars in 2050 [...]; battery storage similarly needs to climb from 0.5 gigawatt hours (GWh) to 12,380 GWh [...] [and] the amount of installed solar PV capacity must rise from 223 gigawatts (GW) to 7122 GW" (IRENA 2018). This massive increase in low-carbon technology puts pressure on the supply chain to produce more products at cheaper rates, which will likely lead to even more cut corners in terms of safety and equity regulations than there are now. We must get ahead of the issue by putting a spotlight on these processes to develop safe, equal, and just regulations throughout all processes of renewable energy technology.

Policy Recommendations

Since its creation, the International Renewable Energy Agency (IRENA) has made a noticeable impact both within global governance institutions and domestically within states around the world. Working as an epistemic organization, IRENA has focused on knowledge sharing between countries and supporting domestic policies that encourage renewable energy projects (Urpelainen 2015; Overland 2018; Mengi-Dincer 2021). As a result, IRENA has established itself, particularly among industrially developing countries, as a valuable and reliable source of knowledge and guidance within the realm of renewable energy policy and projects. However, as the adoption of renewable energy grows, poignant issues surrounding renewable energy projects and technology have also come into the light; particularly issues regarding Indigenous communities' sovereignty (Bohm 2021, Dunlap 2021, Normann 2020, Erickson 2018) and injustices within the upstream and downstream processes of renewable energy technology (Sovacool 2020; Lennon 2018). As the leading international promoter of renewable energy, IRENA has an inherent responsibility to lead the charge in developing policies and regulations to manage and mitigate the aforementioned issues. Currently, the only mention of justice and equality within IRENA's initiatives is found within a new collaborative framework titled "Just & Inclusive Energy Transition," which had its first meeting in May 2021 (IRENA 2021). While the creation of this collaborative framework is certainly a step in the right direction, more must be done to expedite addressing these issues. In the next section, I advocate for the following three policy changes within IRENA: (1) That IRENA create a new initiative dedicated to energy justice within all processes of renewable energy technology; (2) That IRENA actively seek out collaboration with and guidance from Indigenous nations, and work with Indigenous representatives to develop a fair and equitable management framework for conflict between settler states and Indigenous nations; and (3) that IRENA put more focus on supporting de-growth policies within industrialized nations.

First Policy Recommendation – Create an Energy Justice Initiative

The number of human rights violations and power/benefit imbalances within the renewable energy industry is alarming and addressing them needs to be prioritized. To do this, an Energy Justice Initiative could be created within IRENA that is divided into three sections: upstream, midstream, and downstream processes. Focusing on humane and equitable ways of regulating these processes, the initiative will act as a valuable resource for countries focused on mineral extraction, e-waste management, and manufacturing. It will open funding pathways to areas in most need and put a spotlight on exploitative corporations and supply chain actors within the renewable energy industry. The initiative would also provide a platform for research into conflicts between renewable energy projects and Indigenous populations. Much can be learned when looking at several cases with a broader lens, and up to now, most research on green colonialism conflicts has focused on single cases. There are also examples of renewable energy infrastructure being used as a tool of reconciliation between Indigenous nations and settler states through Indigenous ownership of the energy infrastructure (Baxter & Mang-Benza 2021; Campney et al. 2021). In this way, renewable projects act as a tool that strengthens Indigenous sovereignty rather than dismantling it and may open opportunities to solve conflicts elsewhere. Subsequently, there should be a strong Indigenous presence in the structuring and oversight of the initiative, as explained in the second policy recommendation.

Second Policy Recommendation – Heightened Collaboration with Indigenous Nations

Up to now, the voices and opinions of Indigenous communities on renewable energy have been largely ignored or actively silenced, especially within settler states. It is fair to surmise that a large part of the reason for this is because Indigenous narratives challenge the popularly asserted assumption that more renewable energy capacity is always a good thing. Many "Indigenous peoples have denounced how climate change mitigation through quick fixes and large-scale interventions not only dispossesses them of lands and life systems but also limits how we comprehend the current ecological crises" (Normann 2020, 90). However, the relationship between renewable energy advocates and Indigenous voices does not need to be one of contention. After all, Indigenous activists are behind some of the most influential organizations fighting climate change, such as the Idle No More campaign, the Council of Thirteen Indigenous Grandmothers, and the Summer Heat campaign. Furthermore, as described earlier, Traditional Ecological Knowledge (TEK) offered by Indigenous cultures provides a comprehensive blueprint for sustainable living. Scholar Dina Gilio-Whitaker points out that due to Indigenous people's long history of sustainable living and land management, "it may well be that organizing around Native land rights holds the key to successfully transitioning from a fossil-fuel energy infrastructure to one based on sustainable energy" (Gilio-Whitaker 2019, 149). All this is to say that Indigenous peoples and nations will be a powerful ally in the global transition to renewable energy if they are given the respect they deserve and are involved in the planning, construction, and maintenance of renewable energy projects within their ancestral lands.

IRENA can achieve increased Indigenous representation through direct and indirect means. Firstly, IRENA can increase Indigenous representation within the agency itself. Ideally, Indigenous nations would have the same rights as other nation-states and could simply apply for membership within the agency; however, that is not currently the case for Indigenous nations occupied by settler states and acquiring those rights will take years of negotiating and politics. In the meantime, IRENA has other avenues of achieving Indigenous representation within the agency, such as directly hiring Indigenous employees and inviting Indigenous activists and elders to speak to the assembly and advise collaborative frameworks and initiatives within IRENA. Additionally, IRENA can encourage Indigenous representation on individual development projects by leveraging its influence within the policy making and financing sectors. IRENA's support on policies such as more thorough land use assessments, ecological impact surveys, and increased Indigenous collaboration/ownership could help shift the dominant paradigm of renewable development globally, especially if some of these policies are used as a requirement for receiving funding from the Energy Transition Accelerator Financing Platform (a financing platform managed by IRENA).

Third Policy Recommendation – Encourage De-Growth in High Consuming Countries

A successful transition to sustainable living will require more than just increased renewable energy capacity; it requires a lifestyle change for high energy-consuming states like the US and countries within the EU. According to the International Energy Agency (IEA), the annual electricity consumption (in kWh per capita) of the US, Canada, and the collective EU in 2014 was 12,994; 15,588; and 6,022, respectively (IEA 2014). Compared to the consumption of less industrially developed countries like Ethiopia and Vietnam, 69 and 1,424 respectively (IEA 2014), the consumption of the Euro-centric countries is beyond gluttonous. Part of the vision of a sustainable future is better parity of energy access and consumption across the globe, and it is unrealistic to believe that the world can sustain a global average electricity consumption greater than 6,000 kWh per capita per year (the world average in 2014 was 3,131), let alone 15,000 kWh per capita (IEA 2014). Even without taking the issue of energy consumption parity into consideration, the cyclical nature of renewable energy capacity requires a shift in the way high energy-consuming communities use energy; and the higher the energy use, the harder it is to make the necessary shift. Subsequently, in addition to promoting renewable energy, IRENA should also promote policies geared towards degrowth in countries with high energy consumption. As this doesn't necessarily involve all the members of IRENA, it may best be achieved through a collaborative framework where high energy-consuming countries can brainstorm methods to cut down on their energy consumption and agree on annual energy consumption targets for the near future. IRENA could also help with energy audits of countries, finding their highest sources of energy consumption and targeting energy use reduction in those areas for greatest impact.

Conclusion

While renewable energy does have great potential to mitigate climate change issues, a blind faith in the positive impact of renewable technology "ignores the necessity to consider degrowth in those same Enlightened (and colonial) nations; the spread of electrical dependence of people; the mining necessary for it and the different ontologies and ways of life that reject this form of development" (Dunlap 2021, 5). As long as renewable energy technology is manufactured, installed, and disposed of in a way that continues the historic trend of devaluing BIPOC communities, the sustainability of these technologies is put into question. Ignoring the rising energy justice issues surrounding the renewable energy industry would be to repeat mistakes from the past and would surely result in more crises down the line; therefore, the best way forward is to face these issues head-on. The rapid growth and early success of IRENA as an international collaboration organization for promoting renewable energy development foreshadows the influential role IRENA and renewable energy will play within global energy governance in the coming years. The combination of IRENA's existing international collaborative infrastructure and their rising significance within global energy governance puts IRENA in a unique position to lead the world in an energy justice movement—to develop just and sustainable solutions to these issues and disseminate learned practices across all nations. IRENA has done well to recognize this potential by creating the Just and Inclusive Energy Transition collaborative framework but needs to push harder for more centralized action. Specifically, IRENA should create an initiative dedicated to energy justice, actively pursue collaboration with and the integration of Indigenous peoples and nations within IRENA, and focus on promoting reduced energy consumption in high consuming countries. With these policies, IRENA could revolutionize how renewable energy is managed on a global scale.

References

Baxter, J., Mang-Benza, C. 2021. "Not paid to dance at the powwow: Power relations, community benefits, and wind energy in M'Chigeeng First Nation, Ontario, Canada". *Energy Research & Social Science.* 82

Boyles, A.L., Blain, R.B., Rochester, J.R., Avanasi, R., Goldhaber, S.B., McComb, S., Holmgren, S.D., Masten, S.A., Thayer, K.A. 2017. Systematic review of community health impacts of mountaintop removal mining. *Environment International.* Volume 107. 163-172.

BP. 1996 – 2020. Statistical Review of World Energy. BP. https://www.bp.com/en/global/corporate/energy-economics/statistical-review-of-world-energy.html

Bullard, R.D. 1994. Environmental Racism and 'Invisible' Communities. W. Va. L. Rev. 96.

Bullard, R.D. 2005. *The Quest for Environmental Justice: Human Rights and the Politics of Pollution.* Catapult.

Campney, A., Hoicka, C.E., Savic, K. 2021. "Reconciliation through renewable energy? A survey of Indigenous communities, involvement, and peoples in Canada". *Energy Research & Social Science*. 74.

Cook J, Nuccitelli D, Green S A, Richardson M, Winkler B, Painting R, Way R, Jacobs P and Skuce A 2013 Quantifying the consensus on anthropogenic global warming in the scientific literature *Environ. Res. Lett.* 8 024024

Dunlap, A. 2018. "The 'solution' is now the 'problem': wind energy, colonisation and the 'genocide-ecocide nexus' in the Isthmus of Tehuantepec, Oaxaca". *The International Journal of Human Rights*. 22:4, 550-573.

Dunlap, A. 2021. "More wind energy colonialism(s) in Oaxaca? Reasonable findings, unacceptable development". *Energy Research & Social Science*. 82.

Erickson, B. 2018. "Anthropocene futures: Linking colonialism and environmentalism in an age of crisis". *Sage Journals*.

Fixico, D.L. 2021. Documenting Indigenous dispossession. *Science*. 374. 536-537.

Gilio-Whitaker, D. 2019. *As Long as Grass Grows*. Beacon Press, Boston.

Holifield, R. 2001. Defining Environmental Justice and Environmental Racism. *Urban Geography*. 22. 78-90.

Huseman, J., Short, D. 2012. 'A slow industrial genocide': tar sands and the indigenous peoples of northern Alberta. *The International Journal of Human Rights*. 16. 216-237.

IEA. 2014. Statistics © OECD/IEA iea.org/stats/index.asp, subject to iea.org/t&c/termsandconditions

IEA. 1970-2020. Evolution of solar PV module cost by data source. IEA. Paris https://www.iea.org/data-and-statistics/charts/evolution-of-solar-pv-module-cost-by-data-source-1970-2020

IRENA. 2011. *Decisions regarding the work programme and budget for 2011*.

IRENA. 2018. Global Energy Transformation: A Roadmap to 2050. International Renewable Energy Agency, Abu Dhabi.

IRENA. 2021. "Institutional Structure". Accessed on Dec. 11, 2021. https://www.irena.org/About/Institutional-Structure/Assembly

IRENA. 2021. "Vision and Mission". Accessed on Dec. 11, 2021. https://www.irena.org/About/Vision-and-mission

Kaneva, D. 2010-2011. Let's Face Facts, These Mountains Won't Grow Back: Reducing the Environmental Impact of Mountaintop Removal Mining in Appalachia. *Wm. & Mary Envtl. L. & Pol'y*. 35. Rev 931.

LaDuke, W. 1994. Traditional ecological knowledge and environmental futures. *Colo. J. Int'l Envtl. L. & Pol'y, 5,* 127.

Lynas, M, Houlton, B, Perry, S. 2021. "Greater than 99% consensus on human caused climate change in the peer-reviewed scientific literature". *Environ. Res. Lett.* 16 114005

Mengi-Dincer, H, Ediger, V.S., Yesevi, C.G. 2021. "Evaluating the International Renewable Energy Agency through the lens of social constructivism". *Renewable and Sustainable Energy Reviews*. 152.

Normann, S. 2020. "Green colonialism in the Nordic context: Exploring Southern Saami representations of wind energy development". *Journal of Community Psychology*. Wiley Periodicals LLC. 49. 77.

Overland, I, Reischl, G. 2018. "A place in the Sun? IRENA's position in the global energy governance landscape". *International Environmental Agreements*. 18:335-350.

Pulido, L., Sidawi, S., Vos, R.O. 1996. An Archeology of Environmental Racism in Los Angeles. *Urban Geography*. 17:5. 419-439.

Preston, J. 2017. Racial extractivism and white settler colonialism: An examination of the Canadian Tar Sands mega-projects. *Cultural Studies*. 31:2-3. 353-375.

Silva, N. K. 2004. *Aloha Betrayed*. Duke University Press. Durham&London.

Sovacool, B. K., Hook, A., Martiskainen, M., Brock, A., Turnheim, B. 2020. "The decarbonisation divide: Contextualizing landscapes of low-carbon exploitation and toxicity in Africa". *Global Environmental Change*. 60.

Urpelainen, J, Van de Graaf, T. 2015. "The International Renewable Energy Agency: a success story in institutional innovation?". *International Environmental Agreements*. 15:159-177.

Whyte, K.P. 2016. "Indigenous Experience, Environmental Justice, and Settler Colonialism." SSRN: https://papers.ssrn.com/sol3/papers.cfm?abstract_id=2770058

John Hewston, World War II Veteran: Aircraft Gunner, Mechanic, and Supply Sergeant

Michael H. Pazeian (Cal Poly Humboldt)

Keywords: John Hewston, World War II, Wildlife management

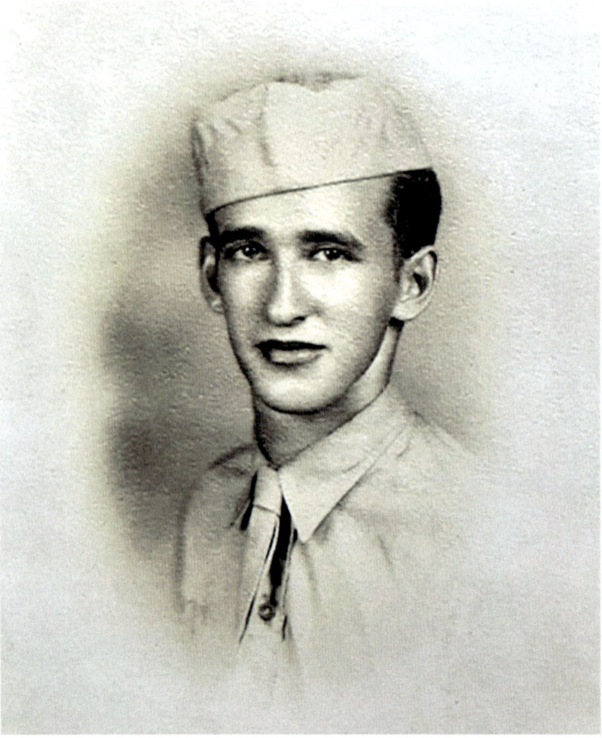

Most of us in and around Arcata, California know John Hewston as a long-time professor at Humboldt State University. He was also known as the birdman. For many years John was with the Wildlife Management Department at HSU, and he spent each Thanksgiving for 25 years coordinating the Bird Count for 12 western states, including Alaska, and compiling the results of observations from hundreds of participants.

His very early years he loved to watch birds. He drew pictures of the birds he saw. As a growing kid birding was unusual, but his friends and family supported him.

This interview was done at John's home in 2019. The focus was his years with the Army Air Corp.

These are John Hewston's words.

I was born in 1923 in Roy, Washington, about an hour south of Tacoma. On December 7th, 1941, I was running a riding academy. I worked and became a horseman. Soon I was teaching others to ride. I was just across the street from McChord Army Air Base. After nine months at the riding academy, I took a job with the Army Corps of Engineers at the base.

John Hewston

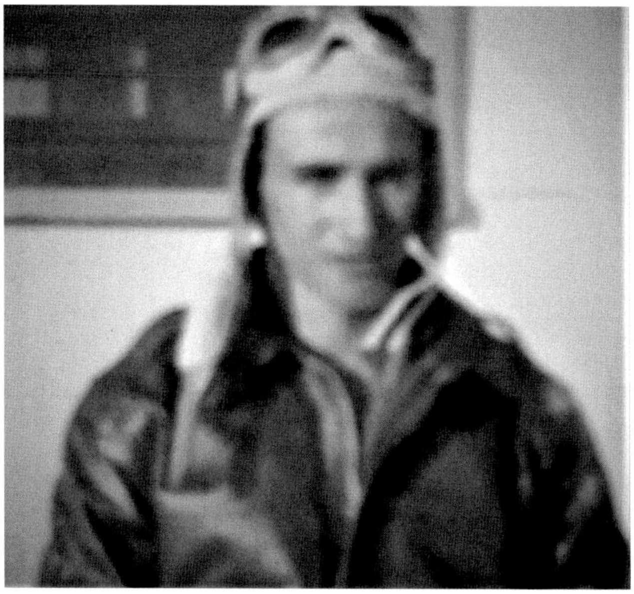

I was working at the base in 1942 when I was drafted. I was young, so the Corps of Engineers had me doing a variety of work on the base: running messages mostly and sometimes janitorial work too. I did that job for nine months. In December of 1942, I was the first teenager drafted in the State of Washington.

Because of the holidays I was told to report on Jan. 15, 1943, at Ft. Lewis, Washington at the induction center. I was sent to Fresno, California for my basic training. On that trip I saw my first palm tree. Most of the guys I was with were much older. We did the same as other basic army units: marching, discipline, guard duty, KP, gun practice, and cleaning the barracks. I was probably there at least 8 weeks. During training there I was really good with weapons. I was a good shot. I picked two things I wanted to do: aerial gunnery school and aircraft mechanic training.

Next was aerial gunnery school at Harlingen Army Airfield, Texas. It was 1943. I was there about eight weeks. In the classroom we studied windage and trajectory—the math behind gunnery. In the field we shot many different kinds of guns—shotguns to learn with moving targets. Mostly 30 and 50 caliber machine guns. At times we shot from planes. I got airsick. The focus of our training was to be up in planes. We went up in single engine and multiple engine planes.

I was an honor grad, top 10, in a class of 300. They asked me to stay on as an instructor. But I turned it down.

I wanted to attend aircraft mechanic's school. I was sent to Keesler Army Air Base near Biloxi, Mississippi. We were in the classroom and working on planes. The B-24 was the major plane. By then I was Pvt. First Class. We also worked on the guns of the plane. Most of the guys with me were from different parts of the country. It took time to get to know them. Many of the guys had to go to gunnery school after we finished mechanic's school. I had already completed it. I wanted to join an outfit going overseas.

My next stop was Clovis Field in eastern New Mexico. I became part of Army Air Corps' operational training unit. Shortly after I arrived, I was on a troop train to Virginia and then to England. But we missed the convoy. We were there in Virginia for a few weeks. I was then we were sent to Georgia because of my aircraft mechanics gunnery training.

I ended up at Chatham Field near Savannah. (Chatham was home to the 8th Air Force's B-24 wings.) Our barracks inside and outside are pictured below. We had time to play some sports. Routines were the same in Mississippi and Georgia.

It was still 1943. I was training new crews on B-24s. Soon I became a specialist in supply. I was in charge of all

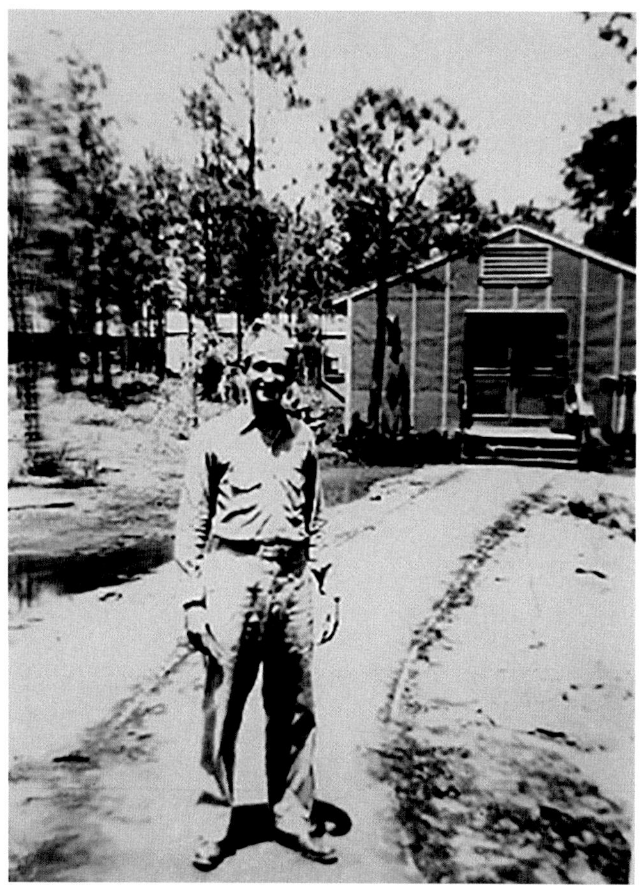

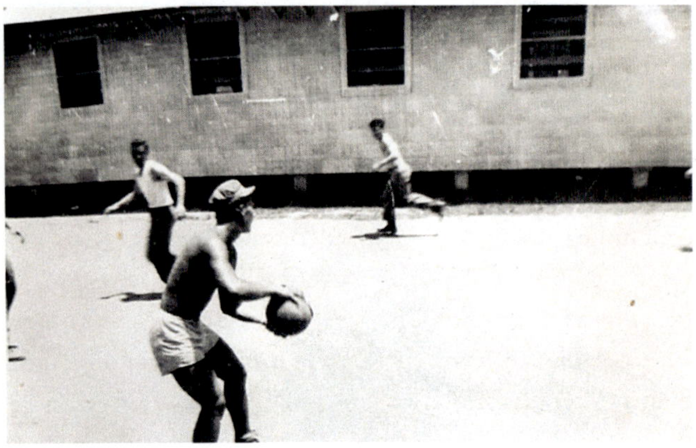

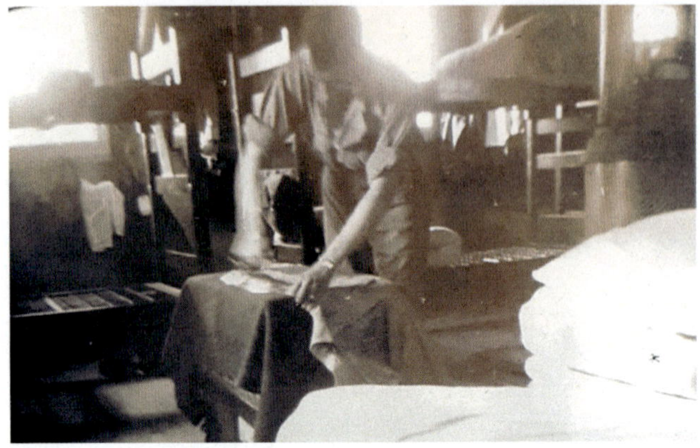

parts for the bombers. Each B-24 had a ground crew. They were responsible for keeping the plane in flying condition. The crew chief would come to me for any and all parts needs. Even parts needed for the guns. Most of the time I was doing my job to supply needed parts. I was now a sergeant.

I knew the plane and all its parts. I tried to make sure we had all the parts needed all the time. I could visualize the part. I knew what parts were inter-changeable. I would be ordering parts from the manufacturer as well. I was in a warehouse type of building on the base. Once in a while, I would go places to pick up parts.

One time my brother came to visit at Chatham. He had just finished his basic training in Florida. Going home to Washington was too long of a trip, so he came to visit me. We went into town and saw a movie.

I worked different shifts. If I had the day shift, I usually went into town in the evening. I had a girlfriend living with her parents in town. A few times I stayed at their house overnight. She and I would good have dinner and see a movie.

One occasion I did fly as part of a crew. The photo above is a practice mission for a squadron I worked with.

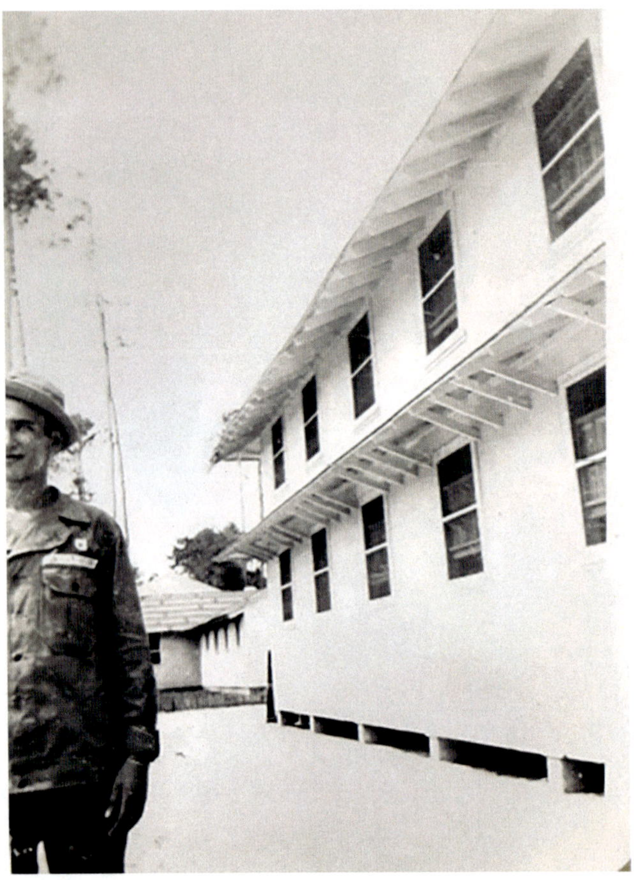

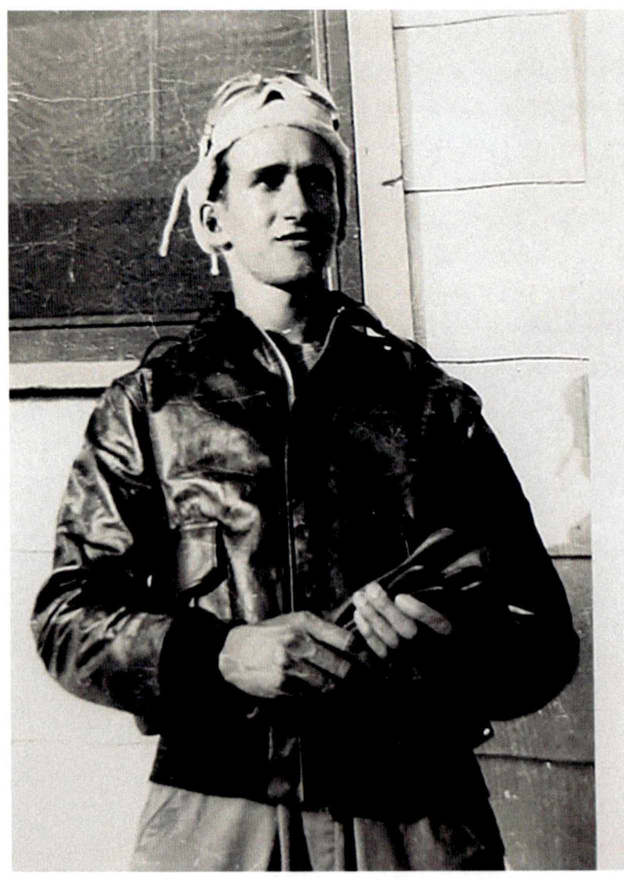

I was declared essential by the Army. I was at Chatham Field until after the war ended.

During the time there I got three furloughs. Three times I took the train home to Washington to visit my family. My younger brother was in the Pacific.

The war ended. I got some training as a military policeman in Florida shortly before I was honorably discharged in early 1946.

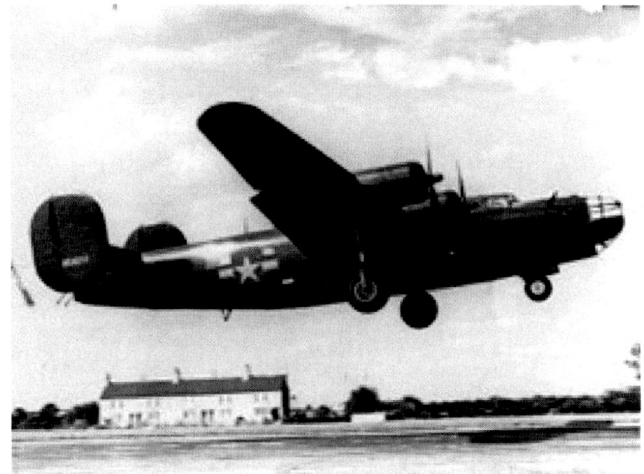

Always interested in wildlife, John began a career with the U.S. Fish and Wildlife Service. He was focus on natural resource management and public relations.

After 20 years, John was hired by Humboldt State University. John's experience with wildlife management was a great addition in 1966.

John still spends lots of time watching for birds from his front porch. He has a house on a wooden lot above the university. At age 98 he and his adult caretakers plan a full day for him. Each of the last three years John has attended the flag ceremony on the Arcata Plaza on November 11.

California State University Systemwide Student Research Competition

The California State University Systemwide Student Research Competition is an annual event that brings together scholars from the 23 campuses of our California State University system. The competition showcases undergraduate and graduate research, scholarship, and creative works by recognizing outstanding student accomplishments.

These Cal Poly Humboldt student articles and videos are reproduced as they were submitted and accepted to the CSU Research Competition.

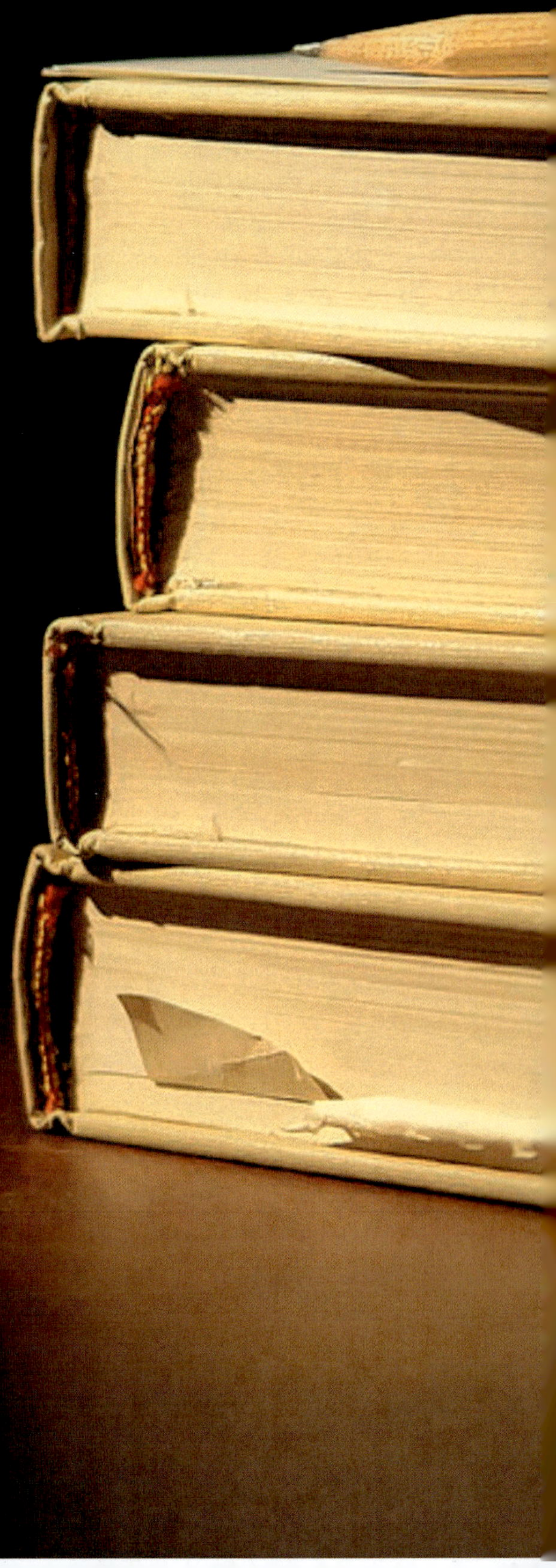

Normative Values of College-Aged Men and Women for the 1.5-Mile Test on a Treadmill for Cardiorespiratory Fitness

Eli Baginski (Cal Poly Humboldt), Eden Marquez (Cal Poly Humboldt), and Skye Choi (Cal Poly Humboldt)

Introduction

University years are a formative time for young adults. Students invest time gaining knowledge in fields of interest, exploring arts and sports, joining clubs, and participating in many other extracurricular activities. It is not surprising that at such a busy time, students do not always take care of their health. In a meta-analysis done by Keating et al. they found that 50% of university students did not meet the American College of Sports Medicine's (ACSM) guidelines for physical activity (Keating et al., 2005). Furthermore, in a review by López-Valenciano et al., the authors found that physical activity levels of university students around the world decreased during the covid pandemic (López-Valenciano et al., 2021). The repercussions of a sedentary lifestyle are far-reaching. Physical activity has been shown to positively affect brain function and cognition (Kramer & Erickson, 2007). A negative correlation has been found between cardiorespiratory fitness and depression, suicide attempts, and self-harm (Grasdalsmoen et al., 2020). In addition to mental health, physical activity affects bodily functioning. Physical activity is one of the chief methods for improving and maintaining cardiorespiratory fitness (CRF). CRF is the ability of the respiratory and circulatory systems to supply oxygen to working muscles (Kenney et al., 2019). Poor CRF has been associated with many diseases from diabetes to heart disease (Al-Mallah et al., 2018). Many of these diseases are among the top ailments that affect people globally (World Health Organization, 2021). Physical activity therefore must be a part of students' weekly routine to help them tackle the difficulty of gaining higher education and maintaining their health at the same time.

With CRF playing such an important role in human health, it is necessary to be able to effectively determine an individual's CRF and whether it meets current recommendation guidelines. Maximal aerobic capacity (VO2 max) is the maximal consumption, distribution, and utilization of oxygen during exhaustive exercise and is the gold standard for categorizing an individual's CRF (Kenney et al., 2019). While VO2 max can be measured by a trained technician using laboratory equipment, most people do not have access to such tests on a regular basis and an easier method for determining their CRF is necessary. Fortunately, many field tests have been developed to allow people to estimate their CRF with little more than a stopwatch and a track to run on. The 1.5-mile test has been reported as one of the most accurate and reliable tests to estimate a person's VO2 max (Mayorga-Vega et al., 2016, ACSM 2013). Using an established equation, the weight and gender of the testee, and run time for the 1.5-mile test, an accurate VO2 max estimate can be produced (Larsen et al., 2002). Furthermore, in an unpublished thesis, Jackson found that 1.5-mile times did not differ significantly ($p = .122$) for either men or women when performed on a track versus a treadmill (Jackson, 2008). To determine a person's level of CRF, an individual can compare their VO2 max score to normative data for their age and gender (ACSM, 2013). Normative data come from cross-sectional studies in which a large group of subjects from a population takes a test. The results of each subject are used to determine percentile ranks that indicate the range of scores within the population. Thus, normative data provide a quick and simple way to assess performance and examine how an individual's score compares to a population (Hoffman, 2006).

The ACSM reports normative data for the 1.5 mile run test, however, this data was not established by sampling a population in a formal study but rather calculated using an adjusted 12-minute run test equation (S. Farrell, personal communication, February 20th, 2021). The ACSM's calculated normative data lack demographic parameters and the generalizability of the data is uncertain. The first purpose of this study is to fill the gap by establishing normative data for the 1.5-mile run test for moderately active university-aged adults. Due to the ease of use, all the 1.5 mile run tests will be performed on treadmills. Secondarily, this study aims to examine any differences between male and female subjects. We hypothesize that male participants will have a lower run time at any given percentile rank than female participants.

Methods

The normative data collected was obtained from a sample of 175 subjects (75 women; 100 men) between 19 and 29 years of age who were recruited for participation in this research. All 1.5-mile run tests were completed on a motorized treadmill (Platinum Club Series Treadmill, Life Fitness, Rosemont, IL). A running distance of 1.5 miles was used to compare results to the normative data as this distance has been previously identified as the minimum distance necessary to estimate cardiorespiratory fitness (Vickers, 2001). In addition, previous research has shown that the 1.5-mile run test had a positive correlation with V⊠O2 max (Gleason et al., 2014, Larsen et al., 2002). Subjects were instructed to run until reaching 1.5-mile with their self-selected stride frequency. The University Institutional Review Board (IRB) approved this study, and each subject signed a written consent form before participating in the study.

Subjects were recruited and tested from December 2015 until May 2019. Subjects who were thoroughly familiar with treadmill ergometry and laboratory procedures, volunteered to participate in the study. Many subjects participated in a club or recreational sports (57.2% of men and 42.8% of women), but not college varsity sports such as soccer, track and field, etc. All subjects regularly participated in moderate or strenuous exercise for a minimum of 3 days per week for a period of at least 4 weeks prior to participation. Moderate physical activity is defined as any form of activity that takes 3.0-5.9 metabolic equivalent of task (METs) to complete, such as brisk walking, dancing, golf, tennis, and volleyball. Vigorous activity is defined as any activity that requires 6 or more METs to complete, such as jogging and running, bicycling, soccer, swimming, or performing heavy lifting (ACSM, 2013). Subjects were screened for cardiovascular and musculoskeletal disease using a medical history questionnaire, an activity questionnaire, and the Physical Activity Readiness Questionnaire (PAR-Q). Inclusion criteria were (a) classification as "Low Risk" according to ACSM's Cardiovascular Disease Risk Factor Assessment (ACSM, 2013), (b) not currently on any type of restricted diet or on any medication, (c) no musculoskeletal conditions or injuries, and no flu or illness during study, (d) non-smokers for at least the past 6 months, and (e) classification as non-obese (BMI < 30 and waist circumference < 102 cm). Based on inclusion criteria, subjects were considered healthy and active. Subjects participated in two separate sessions as part of the study procedures.

All Testing took place in the Humboldt State University Human Performance Lab (HPL). To avoid inter-rater reliability issues, all anthropometric data was collected by the same proctor. During the first session, all subjects completed a consent form, health history form, and had initial measurements (i.e., weight, height, and anthropometrics) taken. Weight (437 Physician's Scale, Detecto, Webb City, MO) and height (Seca 216, Chino, CA) measurements were taken as part of the subject assessments. Body mass index was calculated from height and weight measurements to determine if subjects met inclusion criterion (e). Body density was determined using the 3-site formula using skinfold (Lange Skinfold Calipers, Beta Technology, Santa Cruz, CA) measures at the triceps, suprailiac, and thigh for women and chest, abdominal, and thigh for men (Jackson et al., 1980); ethnic and sex-specific equations were used to calculate the percentage of body fat from body density. Prior to the data collection days, subjects participated in one familiarization session to help them get accustomed to the testing procedures and protocols (i.e., some practices in pacing). Prior to each experimental session, subjects performed a warm-up running for five minutes at their self-selected speed, followed by a dynamic stretching such as leg swing, high knees, etc. Subjects completed 1.5 miles as fast as possible on a motorized treadmill (Platinum Club Series Treadmill, Life Fitness, Rosemont, IL) with a 0% percent grade of incline. Subjects were instructed to run until reaching 1.5-mile with their self-selected stride frequency. Upon test completion, a mandatory cool-down period was enforced. Subjects walked slowly (80 m/min) for about 5 minutes immediately after the run to prevent venous pooling.

Statistical Analyses

The normal distribution of the data will be verified using a Kolmogorov-Smirnov test. Anthropometric data and completion time will be reported as mean ± standard deviation (SD). All data will be analyzed separately to provide percentile values for men and women. The descriptive statistics will be calculated in mean, standard deviation, minimum, and maximum. A t-test for independent means will be used to verify the differences between men and women. Where possible, the data will be compared with normative data for the general population and divided into categories based on physical fitness. All data will be analyzed using GraphPad Prism 9.0 (GraphPad Software, Inc., San Diego, CA) and Microsoft Office Excel for Windows 2016, and significance for all the statistical tests will be set at an alpha level of 0.01.

Results

Anthropometric characteristics and 1.5-mile run test outcomes of the study sample separated by sex are shown in Tables 1 and 2. Most variables were significantly higher in males. The average time for 1.5 miles was 11.7 ± 2.3 minutes for males and 13.3 ± 2.5 minutes for females. The average time for females was 113.3% of the average time for males ($p < 0.01$). The average speed for 1.5 miles was 8.0 ± 1.7 mph for males and 7.0 ± 1.2 mph for females. The average speed for females was 87.3% of the average speed for males ($p < 0.01$). Table 3 contains descriptive statistics and percentile norms for the 1.5-mile run test of speeds and times, respectively.

Discussion

The main objective of this study was to establish sex reference values for the 1.5-mile run test among active healthy adults aged 19-29 years and to compare values between sex. Our data in this population also confirms the common finding of faster speed in men compared to women (ACSM, 2013). Overall, these findings provide useful data that can now be used to interpret individual performance on the 1.5-mile run test for the general population. The results of this study provide normative values for both men and women for the 1.5-mile run test that can be used in a commercial or even home gym setting. Previously, there had not been percentile rankings or other normative data for this protocol that were widely available for general use and interpretation. Now coaches, athletes, and fitness enthusiasts who engage in regular physical activity will not only have access to an aerobic capacity (cardiorespiratory fitness) test that can be administered on a treadmill wherever available and used in place of special equipment and that is sports specific to running-type sports, but they can also interpret their results on their own. This study will provide a convenient method for students to assess their CRF. When improvement is desired, an individual can routinely check progress by referencing the normative data chart as they improve their 1.5-mile time.

Presentation

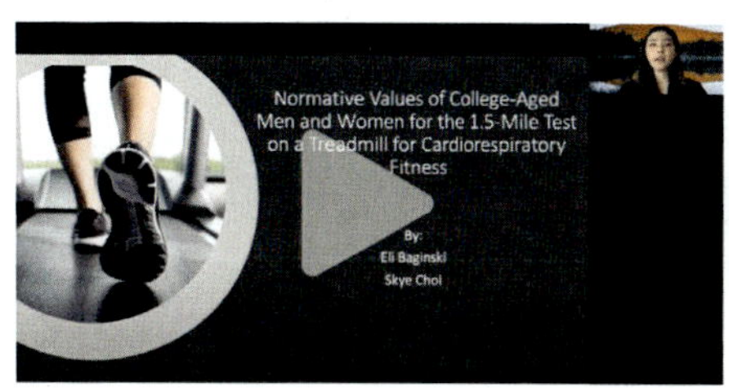

Appendices

Table 1. Characteristics of the study sample by sex. *

Characteristics	All (n = 175)	Men (n = 100)	Women (n = 75)	Difference, p for sex
Age	23 ± 3	23 ±2	23 ± 3	101.2%, p = .015
Body mass (kg)	77 ± 17	84 ± 18	69 ± 10	82.7%, p <. 001
Height (m)	172 ± 10	179 ± 8	168 ± 8	93.8%, p <. 001
Body mass index (kg/m)	25.3 ± 4.6	25.9 ± 5.1	24.5 ± 4.1	94.6%, p <. 001

*Data are shown as mean ± SD *

Table 2. Mean 1.5-mile treadmill run test speed and time for men and women. *

Characteristics	All (n = 175)	Men (n = 100)	Women (n = 75)	Difference, p for sex
Time (minutes)	12.5 ± 2.5	11.7 ± 2.3	13.3 ± 2.5	113.3%, p <. 001
Speed (mph)	7.5 ± 1.5	8.0 ± 1.7	7.0 ± 1.2	87.3%, p <. 001

*Data are shown as mean ± SD *

Table 3. Percentile norms and descriptive statistics for run time and speed for the 1.5-mile treadmill running test for men (n = 100) and women (n = 75).

%ile rank	Speed (mph.) M	Speed (mph.) F	Time (min.) M	Time (min.) F
95	10.7	8.7	08:00	10:17
90	10.0	8.4	08:58	10:45
80	9.1	7.7	09:55	11:43
70	8.8	7.2	10:14	12:31
60	8.1	7.1	11:08	12:39
50	7.9	7.0	11:24	12:43
40	7.5	6.5	12:03	12:49
30	6.9	6.2	13:01	12:52
20	6.4	6.0	14:02	15:05
10	6.0	5.3	15:00	17:00
5	5.9	5.0	15:01	18:02
Mean ± SD	8.0 ± 1.7	7.0 ± 1.2	11:44 ±02:19	13:18 ± 02:28
Minimum	5.8	4.8	15:24	18:42
Maximum	12.1	9.0	07:27	10:02

*Data are shown as mean ± SD

California State University Systemwide Student Research Competition

The California State University Systemwide Student Research Competition is an annual event that brings together scholars from the 23 campuses of our California State University system. The competition showcases undergraduate and graduate research, scholarship, and creative works by recognizing outstanding student accomplishments.

These Cal Poly Humboldt student articles and videos are reproduced as they were submitted and accepted to the CSU Research Competition.

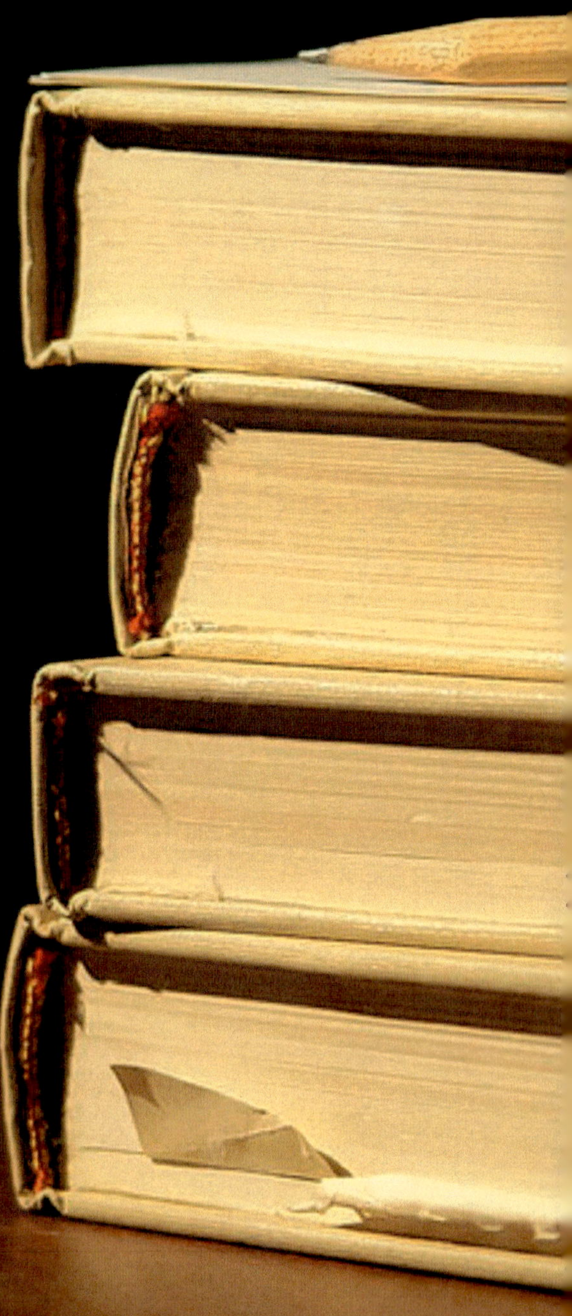

Working Towards Land Return in Goukdi'n: A History of Genocide and a Future of Healing

Carrie Tully (Cal Poly Humboldt)

Since 2009, the city of Arcata, R. H. Emmerson & Sons, and Humboldt State University have collaborated on the transfer of an 884-acre tract of land in Goukdi'n (known locally as Jacoby Creek Forest). The main goals of the collaboration are to prevent fragmentation of the land, protect wildlife, and to support and enhance student research opportunities. My own involvement began just three years ago when I read an article in Humboldt State Now (2018) outlining the partnership and the project. I began this research by examining the history of these parcels. Through document analysis and personal interviews I sought to understand the connection between this land and the Wiyot people. This research provides an interpretation of why relationships to land are not just important, but imperative to healing (Linklater, 2014; Smith, 1999; Dunbar-Ortiz, 2014).

Settler colonialism tried to separate humans from the land and has been effective in some circumstances (Hendlin, 2014). It taught us that land should be thought of as property; something that we own, have the right to control, that is less intelligent than people, and can be contained (Hendlin, 2014; Wolfe, 2006). In order to reconnect with and heal ourselves and our planet, we must return stolen lands to the tribes that are (and have always been) in relationship with them. During my interview with Ted Hernandez, Chairman of the Wiyot Tribe, he stated:

That's why we go back to the land and the healing. You know, we need this healing. We need to bring everything back to balance, so everything can be balanced...You have Goukdi'n, you got Tuluwat, you have Jaroujiji, you know, you have all these special places [that] are part of the puzzle. And if it's missing that piece, that puzzle is not going to be balanced, it's going to fall apart. (Chairman Ted Hernandez, [Interview], June 25, 2020)

Land ownership, domination, and control were the goals of European colonizers since before their feet first touched sand. In fact, the entire system of laws that were enacted since the time of the Homestead Act until the date of the transfer of Goukdi'n to the CSU has been (both intentionally and unintentionally) effective in keeping land out of Native American tribes' hands (Bowden, 2016; Dunbar-Ortiz, 2014; History, 2019; National Parks Service, 2021). Similarly, every step of the way there has been an expressed sense of ownership of Goukdi'n by non-Wiyot people.

With each of these characteristics, it is impossible to deny the deeply embedded traumas that settler colonialism has caused for Indigenous peoples across the world. Historical trauma is the passing down of traumatizing experiences and emotional states generation after generation (Linklater, 2014). Historical trauma refers to "collective and compounding emotional and psychic wounding over time, both over the lifespan and across generations" (Brave Heart-Jordan, in Linklater, 2014, p. 34). What we can do to heal these traumas is address the past and current harms and change our behaviors "in a healing direction" (Linklater, 2014).

Indigenous worldviews are inherently about connection and balance. Traditional healing is "a meaningful step towards attaining a life-in-balance, developing a commitment to self-improvement and healthy relationships with self, others,

Mother Earth, the Cosmos, and the Creator Spirit" (Couture in Linklater, 2014, p. 74). In order to take this step towards balance and healing, it means taking steps toward decolonization. While there are many definitions of decolonization, my favorite version of decolonization comes from Eve Tuck and K. Wayne Yang's Decolonization is not a metaphor (2012, p. 7).

Though the details are not fixed or agreed upon, in our view, decolonization in the settler colonial context must involve the repatriation of land simultaneous to the recognition of how land and relations to land have always already been differently understood and enacted; that is, all of the land, and not just symbolically.

This definition emphasizes that decolonization means to literally return the land, and that decolonization used any other way is just a metaphor and meant to make white people feel better about themselves, which will further settler colonialism.

The entire project has been and will continue to be founded in a version of participatory action research (PAR). While traditional methods of conducting research utilize top-down methods to gather information, naming a 'researcher' and a 'participant', PAR utilizes a horizontal relationship between all parties to the researchers. Participatory action research is not only a means of creating knowledge, and a tool for education, but it also represents the idea that all people create knowledge (Fals-Borda, 2001; Gaventa, 1991). As I move through the project I continue to engage in constructivist grounded research theory, which enables me to seek to realize a social process (Charmaz, 2001) while the data guides the research itself.

Today the Wiyot Tribe has around 350 acres of land as its territory; none of those acres include the redwood forests that were pervasive throughout their ancestral territory. Part of Wiyot ancestral territory is Goukdi'n, incorrectly renamed and referred to locally as Jacoby Creek Forest. Culturally speaking, there are many natural resources that the Wiyot would have come here for, including food, medicine, housing, and basketry materials. The relationship that the Wiyot have with this land is what Dr. Zoe Todd calls a 'storied landscape'. It's one that is reciprocal, meaning that the land cares for the people just as the people care for the land. Therefore, reconnecting the Tribe with Goukdi'n means they are reconnecting with their family, and with their home. Ted Hernandez, Chairman of the Wiyot Tribe states, "the land will always be a part of the Wiyot people and, you know, that's our job is to take care of it, just like we're dealing with Tuluwat now, you know." (Chairman Ted Hernandez, [Interview], June 25, 2020).

The next chapter of Goukdi'n's story began when Mark Andre, former Director of the Environmental Services for the City of Arcata, worked with HSU, the Emmersons, and multiple funders to purchase the parcels. The City, having secured 967 acres of Goukdi'n, kept 83 acres to connect its community forest. The rest of the land was transferred to the California State University (CSU) Board of Trustees, in the care of Humboldt State University free of charge (California State University Board of Trustees, 2018; California State University, 2021; Wikipedia.org, 2021).

Through multiple discussions with the Wiyot Tribe, former College of Natural Resources Dean Oliver, and others, I learned that the Tribe had not been provided seats on either the Faculty Advisory Committee or the Community Advisory Committee. As previously quoted by Dr. Risling Baldy, "The policy or research should explicitly acknowledge the Indigenous cultures and peoples of the area and their continued interaction with biota, landscape, wetlands, or environment" (Risling Baldy, 2013, p. 8) Yet, somehow, so far the Tribe had only heard of this project because of Dr. Risling Baldy and myself, and still had not been approached, even months after the transfer had occurred. In order for there to be progress, we must learn to work together, despite our differences. This particular hurdle was an undertaking by people that I did not suspect to advocate for such a change. In order to work together, we must build respect and trust within ourselves, for one another and for the land. In this case, having love for the land is what has gotten us past this giant step.

Because Goukdi'n is not yet owned by the Wiyot Tribe, the project is not complete. I continue to work with HSU and the Tribe as we move in the direction of this land being returned. It is important to note that Chairman Hernandez and the other Wiyot people who I have been working with have always been explicit about 1) wanting the land back, and 2) that they would continue to collaborate with Humboldt State to provide students and faculty the ability to engage with groundbreaking research in the forest once it is returned. So why go through with a project that will undoubtedly take years, pushback, and an overwhelming abundance of patience? It comes back to healing: healing the past, healing the land, healing the people who were ripped away from their lands, cultures, families, foods...from their very lives. We are all witness to the harms we have caused to Mother Earth. We have violently shoved her out of balance (IPCC, 2021). The return of Goukdi'n to Wiyot peoples could potentially be a precedent setting, internationally recognized action. We could foreground HSU as a leading research institution that not only recognizes Indigenous connections to land, but also supports the ongoing revitalization and resurgence of the Wiyot, not at the expense of HSU, but as an important part of how HSU continues to be a part of this community.

Regardless of the initial oversight of communicating with the Wiyot Tribe regarding Goukdi'n, Humboldt State still has the opportunity to return the land to its rightful owners. As an institution that centers equity, inclusion, and sustainability (Humboldt State University, 2015; Humboldt State University, 2020), HSU can choose to work in partnership with the Wiyot peoples, not only to return stolen land, but also to foreground Indigenous self-determination and sovereignty as essential to research. With the onset of the polytechnic designation at Humboldt State, there has been even more talk about the importance of solidifying partnerships with local tribes, making this the perfect project to align with those goals (HSU, 2021).

Settler colonialism, genocide, US history, historical trauma, healing, and land return are necessarily intrinsically interconnected. Had explorers come to the US and just peacefully coexisted with Indigenous peoples while learning with and growing from one another, I would have written my thesis on something else completely. Had we remained in relationship to place and land, and to our Mother Earth, there would be no need for the rematriation of land. Those are the ways we can imagine otherwise (brown, 2017); a future that is worth living, and one we are worthy of receiving.

Work Cited

1. Andre, M. [Interview]. July 15, 2020. Online.
2. Atkins, M. [Interview]. May 14, 2021. Online.
3. brown, a. m. (2017). *Emergent Strategy*. Chico. AK Press.
4. Brown-Rice, K. (2013). Examining the Theory of Historical Trauma Among Native Americans. *The Professional Counsellor: 3*(3), 117-130.
5. California State University. (2021). Wikipedia. (2021). https://en.wikipedia.org/wiki/California_State_University#Governance.
6. Canter, A. [Interview]. July 7, 2020. Online.
7. Charmaz, K. (2001). *Grounded Theory: Methodology and Theory Construction.*
8. Decolonizing Sustainability: Amplifying Indigenous Perspectives and Transforming Sustainability Discourse [Video]. (2021). Facebook.
9. Dunbar-Ortiz, R. (2014). *An Indigenous Peoples' History of the United States*. Beacon Press. Boston
10. Green, D. [Interview]. June 26, 2020. Online.
11. Hendlin, Y. H. (2014). From Terra Nullius to Terra Communis: Reconsidering Wild Land in an Era of Conservation and Indigenous Rights. *Environmental Philosophy*, 11(2), 141–174. https://www.jstor.org/stable/26169802.
12. Hernandez, T. [Interview]. June 25, 2020. Online.
13. Humboldt State University Polytechnic Prospectus. (2021). Humboldt State University. https://www.humboldt.edu/sites/default/files/hsupolytechnicprospectus.pdf.
14. Humboldt State University Strategic Plan 2015-2020. (2015). Humboldt State University. https://diversity.humboldt.edu/sites/default/files/strategic_plan_2015-2020_final_.pdf.
15. Kauanui, J. K., and Wolfe, P. (2012). Settler Colonialism Then and Now. *Politica & Societa*. ISSN 2240-7901: 235-258.
16. Kimmerer, R. W. (2003). *Gathering Moss: A Natural and Cultural History*. Oregon State University Press. Corvalis.
17. Kimmerer, R. W. (2015). *Braiding Sweetgrass*. Milkweed Editions. Minneapolis.
18. Koch, E. [Interview]. July 16, 2020. Online.
19. Lara-Cooper, K., and Lara, Sr., W. J. (2019). *Ka'm-t'em: A Journey Toward Healing*. Great Oak Press. Pechanga.
20. Linklater, R. (2014). *Decolonizing Trauma Work: Indigenous Stories and Strategies*. Fernwood Publishing Co., Ltd. Halifax.
21. Middleton, B. R. (2011). *Trust in the Land: New directions for Tribal conservation*. Tucson. UA Press.
22. Oliver, D. [Interview]. July 31, 2020. Online.
23. Oliver, D. (July 31, 2020). Jacoby Creek (Goukdi'n) Forest Summary of Activity during HSU's 2019-2020 Academic Year.
24. Project: Humboldt State University/Arcata Jacoby Creek Forest – Expansion. (2018). California Natural Resources Agency. (2021).

25. Resolution 2019-1786. (December 17, 2019). Humboldt County Board of Supervisors Minutes. (2021).
26. Risling Baldy, C. (2018). *We Are Dancing for You.* Seattle. University of Washington Press.
27. Risling Baldy, C. & Tully, C. (2019). Report: Working Towards and For Land Return in Jacoby Creek Forest.
28. Smith, L. T. (1999). *Decolonizing Methodologies: Research and Indigenous Peoples, second edition.* Zed Books, Ltd. London.
29. Tribe History. (2019). Wiyot. (2019). http://wiyot.us/148/Cultural.
30. Tuck, E. and Yang, K. W. (2012). Decolonization is not a metaphor. *Decolonization: Indigeneity, Education & Society,* 1(1): 1-40.
31. Tuluwat Project. (2019). Wiyot. (2019). https://www.wiyot.us/186/Tuluwat-Project.
32. Wilkinson, H. [Interview]. June 30, 2020. Online.
33. Wilson, E. (February 8, 2018). *Returning Stolen Land to Native Tribes, One Step at a Time.* Civil Eats. (2019). https://civileats.com/2018/02/08/returning-stolen-land-to-native-tribes-one-lot-at-a-time/.
34. Wolfe, P. (2006). Settler Colonialism and the elimination of the native. *Journal of Genocide Research.* 8(4). December. 387-409.

Presentation

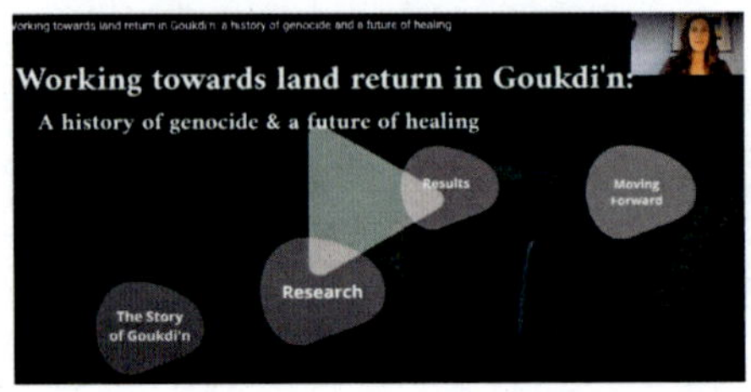

Made in United States
Orlando, FL
27 February 2023

30492206R00035